INDIA CONNECTED

INDIA CONNECTED

HOW THE SMARTPHONE IS TRANSFORMING THE WORLD'S LARGEST DEMOCRACY

Ravi Agrawal

OXFORD
UNIVERSITY PRESS

OXFORD
UNIVERSITY PRESS

Oxford University Press is a department of the University of Oxford.
It furthers the University's objective of excellence in research, scholarship,
and education by publishing worldwide. Oxford is a registered trademark of
Oxford University Press in the UK and in certain other countries

Published in India by
Oxford University Press
2/11 Ground floor, Ansari Road, Daryaganj, New Delhi 110002, India

ISBN-13: 978-0-19-093708-9

Reprinted 2018
Printed in India by Thomson Press (I) Ltd.

For my parents,
O. P. and Neema

CONTENTS

Introduction: The Magic Device 1

PART ONE: Opportunity

1. Google's *Saathi*: Women and the Smartphone 21
2. Hello Macaulay: Fixing Education 43
3. Missed Call: The Smartphone and Job Creation 65

PART TWO: Society

4. Lord Kāmadeva's Digital Bow: Dating and Marriage 89
5. Cyber Sutra: The Internet Is for Porn 115
6. India's iGen: Growing Up with Nomophobia 129

PART THREE: The State

7. Big Brothers: Internet Shutdowns and Internet.org 147
8. The Great Indian Currency Scramble: Digital Money 169

Conclusion: The Everything Device 191

Acknowledgments 207
Notes 211
Selected Bibliography 221
Index 223

INDIA CONNECTED

Introduction

The Magic Device

IN THE DUSTY NORTHWESTERN state of Rajasthan, Phoolwati was visiting a neighboring village on business. She was addressing a small circle of women dressed in sarees. Together, they formed a kaleidoscope of reds, yellows, and pinks. The colors parted obediently when an older woman, in white, pushed her way through the huddle.

"What's going on here?" bellowed the wizened old lady, speaking the rustic Hindi of the region. She pointed at the wiry newcomer, the hub of the commotion. "Who's this?"

All eyes turned to Phoolwati. "I'm here to teach the village women about the internet," she said, as she thrust her hand out, revealing a phone with a large screen of images and text.

She encountered a blank stare.

"In-ter-nate," tried Phoolwati once again, spelling it out phonetically in Hindi. "It's a wonderful thing. You can get all kinds of information and knowledge on it."

The old lady snorted in disdain. "We're all illiterate here, child," she said. "Why are you wasting our time?"

This was a familiar refrain to Phoolwati's internet evangelism. She was prepared.

"Who says you need to read and write to use the internet? Who says you need to know English?" demanded Phoolwati. "This is a *magic* device.

See?" She held up her smartphone and pressed a button. The image of a microphone popped up on the screen. (This might have been more effective had the village women seen a microphone before.)

"Go on. Ask it something," Phoolwati told them. "*Kuchh bhi.* Anything. This has all the answers! You must be curious about something, na?"

The old lady looked on incredulously. She slapped the top of her forehead in an exaggerated show of despair.

Another woman had seen a city cousin toying with a smartphone once. She felt emboldened in the presence of Phoolwati's gadget. "Show us the Taj Mahal!" she exclaimed loudly in Hindi. To instantly summon an image of the country's most famous monument—one that none of them had ever seen—seemed an insurmountable challenge.

But Google understood. The phone came alive; a video appeared on the screen. Phoolwati pressed Play. The strains of a sitar emerged from her smartphone. The scrum around Phoolwati grew tighter, excited, the covered heads of red, yellow, pink, white all pressing against each other, trying to get a better look at her screen. The video showed the city of Agra. It was from the point of view of someone on a *tuk-tuk*, cantering toward the great Indian mausoleum. Then, on foot: some stairs, as the ivory-white marble slowly came into view, step by step, inch by inch. The women giggled at the sight of a white tourist. And then they gasped. Here was the full reveal: glorious, imposing, stunningly beautiful, the great Taj Mahal, one of the seven wonders of the world.

The women broke into applause. They had discovered the magic of the internet.

In New Delhi—a four-hour drive from Phoolwati's village—the city's rich often shop and dine at a ramshackle area known as Khan Market. Inside, the by-lanes are narrow and dusty. Loose electric wires dangle ominously from walls. Stray dogs lie lazily on the pavement. But the shoddy exteriors are deceptive. In the market's shops, one can buy everything from designer clothes to diamonds, exotic fruits to electronics. The restaurants and bars are among the capital's swankiest. In large part because of its location—near Parliament and the homes of the superrich—Khan Market has the most expensive commercial real estate in the country.

Hidden behind the main shopping area is a typically Indian thing: mini Khan Market. This is where the household help, security guards, and drivers of Delhi's elite come to shop. At twenty rupees (thirty cents), a cup of chai costs less than a tenth the price of a Starbucks latte in the main, posh market a few meters away.

Mohit Sadhwani runs a ten-square-foot establishment in mini Khan Market. He is just thirty years old. When unwitting foreigners wander by, Sadhwani exchanges dollars for rupees at extortionate rates. His real business, however, entails selling cheap phones and cellular plans to Delhi's not-so-rich. Sadhwani is a phone wallah—like the chai wallah nearby. Sales are booming.

"No matter how poor, every Indian wants to have a smartphone in his hand," he said, explaining how demand for the internet had made his devices more popular than before. "This is a revolution."

Sadhwani has a point, especially when you compare India to the West. In America, for example, middle-class families began accessing the internet in the 1990s. Most households had personal computers and phone lines, so dial-up internet was a natural step forward. Then came broadband with faster and more reliable speeds. Routers and wireless internet followed soon after. Americans could now imagine the internet on the move: on laptops; across rooms in a connected home, office, or university. Cellular data was invented; people began using email on phones. Then came "smartphones" and wearable technology. All of this was part of a natural evolution.

India's experience has been different. Most Indians don't own PCs or telephone landlines. And now they never will, as they leapfrog straight to the smartphone internet era. For the vast majority of Indians—mostly lower income, rural, or both—internet adoption has not been a steady, Western-style evolution. It is, as Sadhwani puts it, a *revolution*.

Consider the numbers. In 2000, only 20 million Indians had access to the internet. Ten years later, that number grew to 100 million. But 1.1 billion Indians were still offline. Then smartphones and cellular data became mainstream. By 2015, 317 million were online, rising to 462 million in 2017, when three Indians were discovering the internet every second.[1] Women, however, were still left behind: fewer than a third of India's internet users were female.[2] This too will change. By 2020, India's online community is projected to swell past 700 million—about the combined populations of Nigeria, Brazil, and the United States. Nearly a billion Indians are expected to have used the internet by 2025.

This mass discovery of the internet may be a revolution, but it is an unlikely revolution. For all the talk of India as a fast-growing global economic power, it remains a very poor nation. In 2016, per capita income in India was only $1,709 a year.[3] The average Chinese worker made nearly five times as much: $8,123 a year. Mexican? $8,201. American? $57,466.

The question, then, is how the average Indian can afford an Apple iPhone with a voice and data contract that costs an annual $800 in the West. The answer is that he (and increasingly *she*) doesn't need to.

"Apple is only for the rich," said Sadhwani, as he pointed to a store in the main market that sold expensive gadgets. iPhones form just 3 percent of the Indian smartphone market, with Android-based phones accounting for more than 80 percent. "At my shop, people buy the cheaper Indian and Chinese models."

India's Micromax, Lava, Reliance, and Karbonn and China's Xiaomi, Huawei, Gionee, and Oppo produce devices that cost a fraction of the price of Apple products. Reliance's LYF smartphone, for example, retailed for just forty-six dollars in 2017. Pop in a cheap pay-as-you-go SIM card and you're on the web. Discovering the internet has never been so easy.

Indians are getting online in mass numbers because mobile technology has become accessible and affordable. In 2000, a mere 2 percent of Indians were on the internet, in part because less than 3 percent of the population had telephone lines. Meanwhile, 52 percent of Americans were online. By 2020, however, 60 percent of Indians will have access to the internet, as compared with 90 percent of Americans. India will still be poor, but it will have bridged much of the internet gap—almost exclusively by means of cheap smartphones and cellular data.

If the main difference between the shoppers at Khan Market and its mini offshoot is wealth, the ability to speak English runs a close second. Among the British Empire's many colonial legacies is a ruling class of Indians who speak the language of global commerce. Theirs is a small club. Even if one accounts for a basic facility, only about 200 million—or 15 percent—of India's 1.33 billion people understand and speak English. And yet English has long determined one's social standing and likelihood of getting white-collar jobs. Proficiency has also played a part in one's ability to get online. Until a few years ago, most internet users in India were English speakers who naturally tended to be richer, educated people with access to technology. The World Wide Web was virtually nonexistent in Indian scripts. Today, thanks to smartphone software, Hindi, Tamil, Bengali, Kannada, and several other Indian languages are featured on touchpads. Search engines can instantly translate English pages into Malayalam or Gujarati text. For a billion Indians who don't read English, the internet is suddenly accessible. One can go a step further: with voice-activated smartphones, the internet is also available to the people who can't read or write in any language at all.[4] (India's last nationwide count, in 2011, classified 273 million as illiterate.) As Phoolwati and her friends demonstrated, many Indians may not be able to read, but they can speak, listen, and watch.

The smartphone, over time, could be India's great equalizer. It is transforming the country by giving people access to technologies that were until recently the preserve of a rich elite.

To conceive of what is happening in India, we might use an analogy from a century ago: the automobile and the United States of America.

Cars didn't just change how Americans traveled from A to B. They transformed an entire nation. The car created the America we know. It inspired a whole new infrastructure, leading to the building of roads and highways on a mass scale. It launched the vast Interstate Highway System (creating tens of thousands of jobs and boosting the postwar economy).

It defined suburbia. Now families could build picket-fenced homes an hour's drive from the city. The car gave people the freedom to choose where to work and how to get there.

The car led to a dramatic transformation of retail and consumption: it produced the megamall and the supermarket, the factory outlet and the multiplex.

The long weekend. The road trip. The chase. The race. The car was enshrined in Hollywood, on TV, in everyday life. The car fueled teenage dreams. It was romance; it was love; it was sex. It was Freedom. The automobile embodied the American Dream.

The car defined you: it announced whether you were single or married with kids, a city dweller or living in the suburbs. Your politics, your social standing, your income level. Muscle car or sedan? Pickup truck or hatchback? Brand-new or secondhand? Your car made your first impression.

For many Americans the car was their first property, their first truly private—and mobile—space. It was the place many would experience their first kiss.

The internet-enabled smartphone will mean the same for India. To young Indians today, smartphones represent literal and figurative mobility. The smartphone is the embodiment of the new Indian Dream.

Half of all Indians are under the age of twenty-seven and making far less than $2,000 a year.[5] For most, the internet-enabled smartphone will be their first computer—as well as their first private TV screen, their first portable music player, and their first camera. They couldn't afford any of those things before. It brings them the weather, traffic, news, videos, an entire world of knowledge that was previously inaccessible. It helps them travel. In a country with broken infrastructure, it makes life more efficient. It is an escape. In the cramped, tiny homes in which so many live, it is their only privacy. The smartphone will have a profound impact on family life, on all the ways in which Indians live, learn, love, work, and play. Newly connected Indians are already choosing and crafting new destinies. The smartphone offers them a path to a new job and a new economy. It is a low-cost tutor of languages and skills. It eliminates the need to travel five hours to pay an electricity bill. It brings government services—pensions and subsidies—to people who didn't even know they existed. It cuts out the "middleman," for generations a corrupt mainstay of doing anything in India. It allows young men and women access to a dating market that was previously taboo—or controlled by others. Three hundred million children—about the size of America's *total* population—will come of age in an era where smartphones and the internet are truly ubiquitous. They will not even remember a non-connected India.

In a land that has long judged people by caste and tribe, the smartphone is emerging as a new vehicle for self-definition. Your phone announces who you are. Micromax or Samsung? Xiaomi or Spice? Selfie camera or no? Plain vanilla ring tone, Bollywood song, or religious hymn? Boring plastic cover or bling gold?

The internet-enabled smartphone is creating a whole new infrastructure in India, no less radical than the Interstate Highway System. Cashless payment systems like Paytm are encouraging people to experiment with the convenience of digital wallets. India's e-commerce boom is encouraging Amazon and Flipkart to invest in a national supply chain of home deliveries. Uber and its Indian rival Ola are changing how people get around while creating hundreds of thousands of relatively high-paying jobs—until driverless cars become a reality. Maps and GPS systems allow Indians to navigate with a rare confidence. WhatsApp is changing how Indians communicate and do business. Educational apps like Hello English are helping people improve their job prospects. Dating apps like Truly Madly are an alternative to the arranged marriage system. Migrant workers will be able to see their children grow up on Skype.

The internet-enabled smartphone has become one of the most powerful forces in history. But the changes it has brought to the West, for example, have been incremental; it has brought gradual improvements to the lives of people who were already relatively well-off and well connected. The Chinese economy was expanding rapidly before its citizens got online (their internet is also regulated, censored, closed, and dominated by one language). Africa's experience may present some similarities to what has happened in India, but it is, of course, a continent and not a country.

India's experience seems unique: hundreds of millions living in poverty, many of whom are completely illiterate; 800 million under the age of thirty, hungry for opportunities; a country with dramatically unequal access to telephony, electricity, transport, education, health care, and water; a democracy that is home to so many different religions, languages, communities, and cultures. In no other country will access to the internet bring about a change so vast and deep, for so many people, and so quickly. In no other country does it have as much potential to disrupt centuries of tradition and barriers of wealth, language, caste, and gender. These changes were not otherwise going to take place this soon, at least on this scale. As in the West, these shifts will make only an incremental difference for India's elites. But they will have a profound impact on the lives of India's poor—a collection of people large enough to form the world's third-most-populous country. Put together, these internet-driven changes will reshape not only India but, eventually, the global order.

In Kolkata, in the eastern part of the country, the hole-in-the-wall Shaw Auto Parts store is humming with activity. Dozens of customers pop by every hour. But they are not there to buy a wrench or engine oil from Guru Prasad Shaw, the store's founder. Instead they have come to see his two twenty-something sons, Sunil and Sanjay, who are busy punching in numbers on a battered PC. For a couple of years now, they have been running a business helping local workers send money to their villages, buy train tickets, and pay bills. Most of their customers are illiterate. The brothers Shaw charge a small fee to process digital transactions on their behalf.

When I asked them if their business proved that many Indians still weren't equipped to use the internet, Sunil Shaw had a simple answer. "You can give people smartphones all you want. But you can't make them smart."

The Shaws highlight just one of the many hurdles ahead in India's great scramble to digitize. The internet-enabled smartphone is not going to magically fix India's ills. Poverty, for example, is more than just endemic; it is deeply embedded. As I was told by Nandan Nilekani, the cofounder of the tech company Infosys, money is more defining in India than in other cultures. The divide between rich and poor exists everywhere, but nowhere is it more dramatic than in India. Consider the lives of India's elite, the small but fast-growing slice of people who qualify as well-off by global standards. They don't just live apart but can create their own world. Behind their compound walls they have opted out of the worries of the rest of society. Spotty electricity? No problem, they can buy their own generator. Sporadic supply of water? They build their own private water pump. Concerned about pollution? Expensive air purifiers and masks protect their lungs. Worried about a broken education system? Their kids can get private tuition, after which they go abroad to study. More so than in other places, India's rich live parallel to the rest of the country. Money allows them to opt out of public systems they don't trust, which in turn reduces the likelihood of the system ever improving.

The political scientist Francis Fukuyama has argued that "high-trust" societies like Japan and Germany allowed large, shareholder-run companies to flourish because their citizens had developed a belief in collective ethics.[6] Like parts of a giant machine, each group performed its designated function in harmony with a larger system. There were several successful conglomerates in these nations. Fukuyama contrasted Japan and Germany with "low-trust" societies like southern Italy or Taiwan, where businesses were smaller and mostly family run. There were fewer efficiencies of scale. Families tended to be tribal and distrustful of people outside their circles.

Fukuyama didn't classify India, but it would rank low on public trust. For decades, India's biggest conglomerates have been largely family controlled: the Tatas, the Birlas, the Ambanis. When they expanded outside of the immediate family, they often stayed within their own trusted communities: Parsis, Marwaris, Gujaratis. (This has begun to change as India has modernized and companies have listed on stock exchanges—though the impulse can be traced centuries into the past when India existed largely as a collection of princely states.) Political parties in India are similar, with dynastic operations catering to a specific community or interest group. That's why Indians often say about their elections: "You don't cast your vote, you vote your caste."

The result is a lack of true national cooperation. Certain groups flourish at the expense of others, accentuating historic divides between castes and communities. Information and access are traded like commodities. Low trust manifests itself across every facet of Indian life.

I often wondered why Indians would carelessly litter the streets with garbage. The same Indians would take their shoes off when they entered their homes. The home was a shrine to cleanliness and hygiene, but anything outside could be dirtied at will. As V. Raghunathan explained in *Games Indians Play: Why We Are the Way We Are*, Indians do not trust their public works departments to maintain hygiene outdoors. So why bother at all?

If you're a frequent flyer, contrast landing in Tokyo with deplaning in Kolkata. When a plane touches down in Tokyo, travelers will stay buckled in until the aircraft finishes taxiing and comes to a halt. They wait to get up until they are told. The passengers then slowly emerge to pull their well-packed Tumi and Rimowa suitcases out of their luggage bins. They file out in an orderly, precise manner. Why? They trust each other enough to comply with social rules. In Kolkata, the opposite will happen. Travelers expect each other not to comply, to jump queues and barge ahead. They anticipate delays, problems, lost luggage. So as soon as the plane touches down there's a collective scramble: people leap up, grab their bags, and squirm forward. Self-interest trumps compliance because there is a deficit of trust. Indians assume that the system and all its components will let them down.

The problem is systemic. The economist Kaushik Basu once described to me how Indians in New Delhi are habitually late for meetings.[7] The reason, Basu surmised, is economic optimization. Since Indians expect the person they are meeting to arrive late, they plan accordingly. Punctuality is a broken system in India. Yet the same Indian could be in New York and would show up for meetings on time. The reason, Basu said, is that one expects people to be on time in New York, and therefore it is in one's economic interest to be punctual. Put simply, New York has a greater reservoir of social trust than New Delhi.

Lack of trust is also a measure of corruption. The writer Gurcharan Das has pointed out that Indians deal with "birth-to-death corruption."[8] When their child is born, parents pay a bribe to get a birth certificate. And when those parents die, their child pays a bribe to acquire a death certificate, completing a life cycle of corruption. Between these two poles, bribery is a lifelong preoccupation.

A societal trust deficit affects how Western companies adapt to India. The ride-sharing app Uber, for example, is popular in the West in part because it eliminates a cumbersome need to handle cash or cards. As Uber quickly realized when it entered the Indian market, that philosophy doesn't work everywhere. Most Indians don't have bank accounts; most don't trust credit-card payment systems.[9] Indians prefer cash. The local ride-hailing app Ola understood these realities and began offering a cash payment option. It was a success. Uber followed suit. Today, some 70 percent of its rides are paid for in physical notes and coins—an anomaly in its global business.[10] As Ola's founding partner Pranay Jivrajka told me, "Companies need to understand and follow local nuances." India, of course, has an entirely unique set of nuances, with variations for different states and communities.

The trust deficit can be seen in every aspect of how Indians use technology. In the United States and Europe, most people opt for fixed monthly cellular subscriptions and lease their smartphone devices from their service providers. There's a two-way system of trust: customers rely on companies to deliver what they promise, and companies expect their customers to have reliable credit. In India, the opposite is the case. Most Indians buy their smartphones outright in cash from phone wallahs like Mohit Sadhwani; separately, they buy prepaid cellular connections and data packs that they top up as needed, also in cash. They're not locked into a service they might not trust. Indians often sport smartphones with not one but two SIM card slots: they switch between service providers depending on cost and quality, constantly monitoring and taking advantage of special offers and discounts.

Indians find unique ways to avoid paying big companies for downloading content. In corner shops across the country, a service called side-loading transfers apps, games, music, and movies onto a smartphone using a physical cable. Side-loading—essentially mass piracy—is commonly accepted across India. These transactions are a conspicuous result of a lack of trust. It's a small win for the little guy, sticking it to a big corporation. Smartphones won't fix these collective trust deficits. If anything, the opposite will happen. India's unique conditions will shape how technology is used.

Trust has also been eroded by India's broken infrastructure. The World Economic Forum's Global Competitiveness Index shows that despite recent improvements, India ranks 66th in the world for its basic infrastructure in roads, bridges, and electricity. It ranks 91st in the world for its institutions of health care and primary education, and 75th on higher education and training.[11]

Other indices bode no better. Transparency International ranks India's corruption 81st in the world.[12] Women have dismal prospects. According to the World Economic Forum again, India's women rank 139th in the world for economic opportunities.[13] The caste system is formally abolished, and yet it has endured as a means of power projection: India's National Crime Records Bureau notes that in 2016 there were 40,801 crimes by upper castes against lower castes.[14]

To expect that internet-enabled smartphones will whisk away these problems is dangerous. Some argue they might make them worse. Terrorists are today better equipped to communicate, recruit, and radicalize online. Fake news is an old phenomenon, yet it has been truly weaponized with the widespread use—and abuse—of social media. The same goes for "trolling," hate crimes, and cybercrimes. Most Indians aren't sensitized to technology and therefore may be more susceptible. "Death by selfie" has become a real category of accident these days, as people seek ever more adventurous ways to photograph themselves: by the sea, atop tall buildings, hanging on train railings. Data show 60 percent of the world's selfie-related fatalities occur in India.[15] A decade ago, most Indians had no way of accessing pornography; today, Indians make up the world's third-largest viewership for online smut, behind only the United States and the United Kingdom.[16] The most-Googled person in India is . . . a former porn star. Families are being forced to contend with smartphone addiction, a problem so pervasive that "de-addiction" centers have sprung up in Indian cities. Doctors are warning of a new disease: nomophobia, or no-mobile phobia, the fear of being away from one's phone.

Smartphones are at the heart of a cat-and-mouse tussle between the people and the powerful. It has become routine for Indian intelligence to tap the phones of prominent journalists and academics. Most resort to apps like Signal and WhatsApp—which offer end-to-end encryption—to keep their messages private.

There's also the question of the role of government. Some argue the police should have the ability to shut down the internet—just as they can close off a road or highway—during outbreaks of communal violence. Rights activists counter that no government should be trusted with such power. Either way, India already has the dubious record of ordering the world's highest number of internet shutdowns; only Syria and Iraq come close. In the conflict-ridden state of Kashmir, for example, citizens went without the internet for several months in 2016, causing hundreds of millions of dollars in business losses.

Indians may lack faith in their systems, but in perhaps the greatest national exercise in trust ever attempted, a billion citizens have handed the government their biometric data—fingerprints and retinal scans—to create new digital identification cards. The program is called Aadhaar, and it has built the world's biggest biometric database, one that may form the basis of identification, banking, subsidies, voting, health care, and more. Perhaps it's a blind trust, because much could go wrong with it (and in several cases already has). Indians are engaged in a vigorous debate about privacy and data security. But by and large, the nation seems to be putting its faith in better times ahead. Digital times. This is where the smartphone's influence will be greatest.

The Indian smartphone story begins before smartphones, before the internet even, and is marked by three distinct moments. The first occurred in 1991, when India was on the brink of economic ruin. New Delhi was weeks away from defaulting on its international bills. Gold was flown to London and Switzerland as collateral for a loan from the International Monetary Fund. It was a symbolic occasion of national shame, like a mother pawning her jewels to buy food for the next week.

Some prefer to remember this as the beginning of India's economic turnaround. A looming disaster led to bold reforms, flinging open a window to foreign investment. In those days the state of West Bengal—of which Kolkata is the capital—was ruled by Jyoti Basu, the world's longest-serving democratically elected Communist Party leader. Even in Kolkata, which ideologically and defiantly resisted change, there was Pepsi and Coca-Cola, Levi's jeans and Benetton shirts. A Wimpy's and then a McDonald's arrived to compete with the older restaurants in the city. Protectionism was defeated; the world was being allowed to trickle back in.

The year 1991 enabled a second agent of change: cable TV. Finally, people had options that went beyond the state broadcaster Doordarshan. (The Hindi word *doordarshan* means, quite literally, "viewing things from far away." But not *that* far away, because all the channel would bring was Indian movies, staid news bulletins, nationalistic commercials, and dramatizations of Hindu epics.)

Cable TV exploded the old menu. Now middle-class Indian teens had MTV and *Baywatch* in the comfort of their homes. They had the BBC and CNN. Cable TV brought the world to the Indian living room. Even in slums and remote villages, large groups of people would crowd around the one TV and watch everything from live cricket matches to blockbuster movies.

The mid-1990s saw young Indians shift their horizons. They began to dream bigger, bolder, more colorful dreams; their sense of fatalism eroded; the world became smaller. Thanks to cable TV and its many cultural influences, Indians began to look more to America than to England for inspiration. The American Dream resonated with young Indians, contrasting sharply with the more hierarchical English notions of class and society that had long pervaded Indian culture.

A few years later, in the late 1990s, came the third wave of changes, one that allowed the world to encroach still further into Indian life: mobile phones and dial-up internet. For millions—still only a tiny fraction of the population, of course—they brought choices, options, efficiency, privacy. More consequentially, they set up India's eventual path toward internet-enabled smartphones.

In 2001, when I was eighteen, I left India to study in the United States. I returned often to see family, but for short stints. With each passing visit it seemed India was building up into something bigger, the way each new page of a Lego manual reveals another layer of construction, except with no clear end in sight. India's cities seemed to become bolder, and its citizens more modern and self-assured. A movement of cultural and linguistic reclamations began, part of a renewed pride in the notion of India: the anglicized Calcutta became Kolkata; Bombay became Mumbai; Madras, Chennai; Bangalore, Bengaluru; and so on. A new aggression and confidence characterized how Indians played sport, practiced diplomacy, conducted business, won beauty pageants, and saw their place in the world. Nationalism began to creep into Indian politics. Unimaginable wealth was being created—and spent—in India's biggest cities. India remained poor, but the trajectory seemed to curve unmistakably upward.

In 2014, I returned to India as a journalist reporting and managing CNN's coverage of South Asia. The sense of exuberance so many Indians seemed to feel about their futures was now unmistakable: rising income levels, the country's youthful demographics, opportunities for professional growth. Young Indians seemed idealistic and upbeat, far more than I remembered being at their age, and far more so, at least in my view, than their counterparts in Europe or America. Much of this exuberance felt irrational to me; India was still poor, underdeveloped, and struggling to create jobs for its young men and women. But hope is relative, and for perhaps the first time in centuries young Indians, in thousands of villages and small towns, were beginning to feel unbound by their fates, by their castes, families, and traditions. They believed that their lives would be better than those of their parents.

There were many factors at play here, but the one thing that encapsulated the moment was the smartphone. India and Indians had already embraced basic cellular phones—what is now euphemistically known as a feature phone or a keypad phone. By 2014, however, everyone seemed poised to jump to touchscreens and the promise of all the freedoms the internet could bring.

This time, it truly was for *everyone*, not just the urban elite or the English educated. Or so we were told to believe. Everywhere I looked in India—roadside billboards, TV commercials, Bollywood movies, chai stalls—in cities and in villages, the smartphone was animating a new energy and optimism. Internet and phone companies dominated advertising, sponsoring everything from the Indian cricket team to concerts and TV entertainment. And in their advertisements, the smartphone symbolized hope, progress, equality, freedom, aspiration.

A fair bit of this was propaganda, of course. Millions who couldn't afford smartphones were still purchasing the cheaper keypad options. The quality of cellular service—with slow data speeds and frequent call drops—remained poor. Older Indians struggled to keep up with technology and felt left behind. There has been little public debate about how to prepare people for the world of smartphones and the internet.

In 1853 Karl Marx wrote that modern industry in India—built on the back of the railways—would lead to the end of hereditary labor and the caste system.[17] Similar predictions have been made about electricity, TV, the basic cellular phone, and the growth of the middle class. India has had several transitional moments that were not transformative.

One reason the smartphone may succeed is because it is working alongside other transformations. A confluence of trends has made India ripe for the changes that technology can unleash. On the one hand, advances in science and globalization have led to the creation of computing tools that are powerful, small, and cheap; on the other hand, Indian companies have grown to a point where their spending power can realize mass efficiencies of scale. India's Reliance Industries touched a market capitalization of $90 billion in 2017, a wealth that enabled it to launch a free nationwide 4G voice and data service called Jio (the Hindi word *jio* translates as "live life"). One hundred million people subscribed within six months of the service's introduction. Meanwhile, per capita income in India nearly quadrupled between 2000 and 2016.[18] As Indians get richer they will consume and spend more, much of it online and on their newly acquired smartphones.

The potential for these changes has sparked a corporate gold rush to India. Apple's Tim Cook has said he is "very, very bullish and very, very optimistic about India."[19] Amazon's Jeff Bezos has invested $5 billion to capture an e-commerce market that HSBC bullishly claims will grow in size from $21 billion in 2016 to $420 billion in 2025.[20] Facebook and WhatsApp both have more users in India than in any other country—including the United States.[21] One would be hard-pressed to find a giant multinational unwilling to bet on the Indian market. China is pouring in money. At the end of 2017, Xiaomi overtook Samsung to become the top seller of smartphones in India. Meanwhile, Paytm, BigBasket, and Hike—the country's biggest online payments company, grocer, and messaging company, respectively—each count a Chinese firm as its biggest investor.[22]

For its part, India is betting heavily on this potential—and there are risks. What will this new device, this new wave, bring? What will be gained, what lost? How will disruptions shape the things that define India, with its ancient traditions, culture, and society? How does one reconcile the age of omnipresent technology with a civilization that gave birth to yoga and four world religions?[23]

This book attempts to answer those questions. The chapters ahead will focus on the aspects of everyday Indian life that seem poised for disruption by widespread internet use: the status of women, education, jobs, dating, marriage, family life, commerce, governance. Phoolwati believes the internet-enabled smartphone truly is a magic device, one that will ensure her daughters have a better life than she did. Some men I encountered want to prevent women like Phoolwati and her daughters from using smartphones, but one expects such repression to be nearing its end. Abdul Wahid, a science tutor, considers smartphones to be a distraction for his students. And yet *he* mastered English—attaining the confidence to run a tutoring company—with the help of a smartphone app. Sudhir Prasad, an Uber driver, has seen his income rise and then plummet because of a company algorithm; meanwhile tens of thousands are still lining up for jobs like his. Simran Arora finds the love of her life on a smartphone app but discovers that Indian families remain stubborn about culture and tradition. Fourteen-year-old Saikat Sinha is addicted to his smartphone. A growing body of research shows teens like him could suffer from depression. Later in the book I explore the devastating—even deadly—impacts of fake news; of how people in certain parts of the country have been systematically cut off from the internet; of how technological advances will leave millions behind. Change can be painful and uneven.

Ultimately, though, access to the internet will help more people than it will hurt. Will the smartphone build trust? The magic device shows the most promise of democratizing things; the trust has to build off that.

I have lived exactly half of my life with and half without the internet. I have also spent half of my life in the West and the other half in India. My hope is that this will help me to explore India's road ahead. The internet-enabled smartphone represents the greatest national upheaval since Partition in 1947. The mass proliferation of these devices means India is being rewired: literally, physically, socially, culturally. The smartphone is at once the greatest challenge to the nation's stability and also a means to build ties.

Any book about such a vast country is bound to generalize. As many have said before me, writing about India is akin to writing about Europe or Africa. India cradles so many different cultures, histories, languages, cuisines, religions, traditions. And yet, as the writer R. K. Narayan said in 1961, "India will go on."[24]

As much as possible, I avoided the use of translators; I speak Hindi-Urdu, Bengali, and some Marwari, which helped in getting my subjects to open up.

I owe a deep debt of gratitude to those who have shared with me details about their lives. Their insights gave me a glimpse of where this great and bewildering nation is headed and how smartphones are shaping the country's destiny. This book is their chronicle: one more chapter in a story that "will go on."

PART ONE

Opportunity

I

Google's *Saathi*

Women and the Smartphone

AS NIGHT FELL in the tiny hamlet of Nangli Jamawat, a light glowed within one section of a small two-roomed home, like a soft beacon. The house was rectangular, built of brick with gray cement coarsely patted over its surface.

Inside the main room were two wooden beds, side by side. On one, two young children were sleeping soundly, neatly curled into cashews. On the other lay Satish, on his back, as his wife, Phoolwati, readied to join him. A cool desert zephyr had seeped under the door and into the room; human warmth was a welcome relief.

In a bare corner, on the cement floor, a thin white wire coiled to a rectangular object that was the source of the light. Phoolwati had set her smartphone down to charge. As she lay down, before she could shut her eyes, the rectangle of light faded into a black mirror. Room, home, fields, village became night.

Phoolwati had been the first to show her fellow villagers a smartphone.

"Want to see a miracle?" she asked a group of women sitting together and tossing rice on wide bamboo sieves, separating the grain from its residual husk. "There's this new thing called Google—want to see?"

"Goo-gull? What's that, a game?" replied one of the women blandly, bored, barely looking up as she kept tossing grain. Her name was Chameli.

"No, no, it's a really useful thing. It comes on the mobile," said Phoolwati, pointing to her smartphone. "If you want to learn about the best seeds to plant, you ask this thing. If you want to know how to get government money to build a toilet, you ask Google. It has all the answers in this world."

Chameli stopped to look up at Phoolwati. She raised an eyebrow theatrically, as if to ask: "Don't you have anything better to do, woman?"

Phoolwati was not deterred. "It has the *Hanuman Chaleesa* also, set to beautiful music," she tried again, changing tack to matters of the divine. The tossing paused. The name of their chosen deity, the monkey-god Hanuman, made her audience sit up and take note. Here was the opening. Phoolwati had succeeded in her saleswoman's pitch. She sat down, sandwiching herself between Chameli and her cohorts. "I told you, na, anything you need—*anything*—Google-*mein-dal-do*." Put-it-in-the-Google, as if it were a cauldron of spells, bubbling with answers. "How much rain will fall this year... how to make softer rotis... how to teach your children... this has everything!"

Nangli Jamawat was part of a cluster of villages tucked away near Alwar in the northwestern region of Rajasthan, India's largest state. It had become a common sight in the area to catch Phoolwati on her bicycle: head down, saree hitched up, legs pedaling furiously. At the back of her ride was a toolbox with the words "Internet Saathi" printed in the Hindi Devanagari script.

Internet Saathi was a nationwide program launched in 2014 by Google and India's Tata Trusts to get rural women online (and, presumably, "putting things in the Google"). Phoolwati was a top *saathi*—meaning "companion"—tasked with showing village women the magic of the internet.

The need for this service was clear. In 2014, only one out of ten women in rural India had ever been online. There were many factors contributing to this worrying statistic: poverty, illiteracy, misogyny, to name just a few. The conceit of the program was that the saathis would succeed where society and the state had failed. Trainers like Phoolwati, as peers who spoke the vernacular, would be able to relate to local nuances. But the saathis themselves needed considerable training and assistance. Google therefore partnered with local NGOs to teach recruits the basics about smartphones and internet searches. Once they were ready, they were handed the tools of their trade: a blue bicycle, internet-enabled smartphones and tablets, and a monthly stipend of about forty-five dollars. And then they were deployed, every day to a new community, like zealous missionaries of technology. Often, the saathis were also given an assortment of portable power banks to tide them over when they had no access to electricity.

Google's India chief, Rajan Anandan, calls India's rural women "the final frontier" in the race to get India's nonconnected online. "There will be big changes in our society when the next 300 million Indians—the middle classes—get on the internet," he told me. "But when the final 400 million get connected, most of whom are rural women, *that will be transformative.*"

The task is immense. While on the one hand India has produced giant information technology companies and armies of coders—used by corporations around the world—hundreds of villages remain without electricity, and thousands without telephone lines and computers. As an additional deterrent, vast numbers of Indians cannot read or write. The country's last official survey, a census conducted in 2011, recorded 273 million people over the age of seven who were illiterate.[1] Women were worse off: while the national rate of literacy for males was 82 percent, the rate fell to 66 percent for females. Some Indian states recorded medieval numbers. Rajasthan ranked dead last with a literacy rate of 53 percent. In Nangli Jamawat, which doesn't have an official rate, Phoolwati reckoned just one in five of the adult women she encountered was literate.

How do you get online when you can't read? Voice technology is already a game changer and has accompanied—even accelerated—the proliferation of cheap smartphones and cellular towers. Indians too old to become literate now have the internet within their grasp just by pressing a button and speaking. And Google is truly multilingual: it can understand everything from Bengali to Hindi, Marathi, Tamil, Telugu, and more.

Phoolwati, sensing that she had opened a door to converting her new audience, decided to push ahead.

"Ask Google," she instructed Chameli, as she held up her smartphone. "Tell it to play *Hanuman Chaleesa.*"

Smiling shyly, Chameli gave it a try: "*Hanuman Chaleesa lagao!*" she said, in her Rajasthani dialect of Hindi.

Google hesitated. The search results displayed the country Senegal, six thousand miles away from Rajasthan. The women squinted at the screen, brows furrowed, as they examined the green-yellow-red flag.

"No, no. Again," urged Phoolwati, holding her phone closer to Chameli's face. "Say it clearly, na!"

This time, a video popped up. Phoolwati hit Play. Cellular data is often spotty outside India's cities, so it took a moment for the video to buffer. But then the sounds of a tabla and a sitar emerged from the device. And then a voice, singing a rendition of the popular Hindu hymn.

"*Arre vah!*" Chameli chortled, clapping with the happiness of a child who had discovered a new toy.

Skepticism transformed to enthusiasm was routine for Phoolwati. Her trainers had given her a list of things that her audience would find useful: recipes for common Indian snacks, weather alerts, government pension programs, religious hymns, and videos of famous sights around India (including, as we saw earlier, the Taj Mahal). For children, there were videos to teach basic math, Indian languages, and English.

Phoolwati was one of the few literate women in her village, but she had only rudimentary skills. She knew how to type out her name in both Hindi and English on Google, but she couldn't speak or understand any English. Phoolwati liked showing people a picture of herself on the Internet Saathi website. She did not grow up with photographs.

"OK, now let me show you how to make samosas the right way. Come, look here!" Phoolwati went on, in a singsong tone she adopted when she was teaching.

Over the course of eighteen months, Phoolwati had trained fifteen hundred women—averaging more than twenty a week. Her main problem was continuity. Once she showed a group of women how to use the internet, explaining how it could be useful to them, she would leave not knowing if they would retain her lessons. There was no guarantee they would be able to get their own smartphones. In many cases, their husbands either couldn't afford or wouldn't allow it.

"What to do," Phoolwati confided to me. "The men come first here. I find the women and I teach, teach, teach. Then they send me to a different village. And many of the women forget what I taught them."

A local NGO called Ibtada—the Urdu word for "beginning"—would decide where Alwar's saathis were deployed. Manoj Kumar Sahni, who trained and mentored the local saathis at Ibtada, would rather have breadth than depth when it comes to spreading the internet. "India is a vast country," he told me. "I have to meet my targets. Even a little knowledge is better than nothing."

Sahni himself only discovered the internet a few years back, on a smartphone. It changed everything for him. He followed the news, got weather alerts, and stayed in touch with his cousins in a far-off village. He discovered a passion for photography. Sahni loved to teach but was convinced that the fastest way to get more Indians online was by getting women to take charge. "I can teach ten women a month. But those ten women can teach ten thousand others that I can never reach," he said. "Only a local woman can really understand what other local women need. The best teaching is by peers and friends, not by professors."

At the start of 2018 there were thirty thousand saathis operating in twelve of India's twenty-nine states. Google said the program had spread to 110,000 villages and benefited 12 million women. The ratio of rural women who had used the internet—even if they weren't regular users—had grown to three out of every ten. "But it is a drop in the ocean," lamented Sahni. "We have so much more to go. We can convert them, we can teach them, but ultimately they have to buy their own smartphones, na? They need family help for this."

Phoolwati is hopeful changes are on the way. The internet has given her a confidence in the unexpected. "Any questions we have, we ask. And it goes straight from our lips to Google's ears," said Phoolwati. "It's like a miracle. That's why I call this phone my magic device."

Indian villages rarely make national headlines unless it's *really* bad news. A gruesome gang rape, a train crash, a wave of farmer suicides, a so-called honor killing: those are the stories that tend to drag reporters and TV crews from the big cities out into India's hinterland.

The village of Suraj in Mehsana, Gujarat—nearly five hundred miles southwest of Phoolwati's Nangli Jamawat—had its turn in the spotlight in February of 2016. Local media reported that the head of the local *panchayat*—a council of village elders—had issued a decree stating girls under the age of eighteen would no longer be allowed to use smartphones.

The Press Trust of India quoted the *sarpanch*, the panchayat chief, as saying the council had reached a unanimous decision to impose a fine of 2,100 rupees on girls who violated the new rule.[2]

"Everyone knows what happens in today's world due to mobile phones," sarpanch Devshi Vankar was quoted as saying. "This is an era of WhatsApp, where people secretly talk with each other. We have to save girls from those who acquire their number and harass them."

The Press Trust of India report continued to cite Vankar: "School girls are getting distracted due to mobile phones because they play games on it and do not concentrate on their studies."

There was no mention of boys.

The story quickly got national attention. Gujarat, after all, was the home state of India's Prime Minister Narendra Modi, a politician who had promised on the campaign trail two years earlier that he would ensure every Indian had a smartphone in his—or her—hand. Modi had also launched, with much fanfare, a program called Digital India, which aimed to cut through bureaucracy and put essential government services on the internet. The smartphone, of course, was the only way most Indians were going to get online.

Suddenly, a sleepy little village of fifteen hundred was flung into the national consciousness as a poster child of how Old India was colliding with Future India.

It turned out the debate at the heart of the panchayat ruling in Suraj, unbeknownst to the villagers, had recently been highlighted on film. The 2015 movie *Parched* is about the women of an unnamed desert village somewhere in northern India. The women struggle against social evils, alcoholic husbands, and young sons who have warped views of the other sex. One of the early scenes in the movie shows a panchayat meeting where the head of the village women is given a chance to address the gathering.

> "Namaste," she says, getting up to address the group. The women are all sitting cross-legged on the ground together, segregated a few meters from the men.
>
> "The women have a request," she begins. "The trucking contracts keep our men away for weeks, while we women are trapped inside our homes."
>
> She inches tentatively toward her point.
>
> "We are the last village without a TV."
>
> The shot cuts to the panchayat men, who are now looking at each other in mock anger and surprise at her temerity.

The scene cuts back to the woman, as she continues to implore. "TV would entertain the men too. Then they wouldn't need to visit the dance company!"

Some of the men begin to laugh, knowingly. (The traveling dance company plays an important part in the movie. One of the main characters is a woman called Bijli—"electricity"—who puts on dance performances for the men, doubling later at night as a high-end prostitute.)

Now it is the sarpanch's turn to respond.

"We already regret giving in to your last request," he says. "You demanded mobile phones to keep track of your husbands. We relented. But what happened? Shiva's daughter eloped with a man she met on the mobile!"

The panchayat men murmur their approval.

One of the others calls out: "You know, in the next village, girls who watch TV are starting to wear jeans!"

The men jeer.

And then another one hoots: "Hear this, brothers! Next thing we'll know these women will be driving trucks as well! Trrrrr *trrrr*," he finishes, acting out the sound of a truck.

The men roar in derision.

When I inquired about the real-life panchayat ruling in Suraj, I was struck by some of the similarities with the scene from *Parched*. In Suraj too, I was told, there had been incidents where village girls had eloped with men from neighboring areas; the blame was invariably placed on the cell phone as an enabler of these illicit love stories.

I wanted to investigate further. Many of the media reports I read seemed to have datelines from Ahmedabad, the nearest big city in Gujarat, which suggested to me that few reporters had visited Suraj itself.

When I reached Devshi Vankar on the phone days later, it turned out he was no longer the sarpanch, perhaps as a result of the media furor. He denied that any sort of ban on phones had ever taken place.

"It's all wrong news," he told me. "Were you here? Was anyone here? No. There was no such ruling. Why would we ban anything? All false! All the women have what they need."

Vankar hung up.

I called again and asked if he would meet me. He refused.

I didn't write about Suraj at the time, but I made a mental note to visit when I could.

Many months later, my chance arrived. It was the same week as Prime Minister Modi's Vibrant Gujarat summit in the capital, Gandhinagar, an hour's drive from the city of Ahmedabad.

Modi's supporters often refer to his twelve years as the state's chief minister—right before he became PM—as the "Gujarat Miracle," a period of sustained growth with the mass construction of highways, roads, and canals and a surge in trade and business investment. Modi's critics, however, point to worrying statistics on human development: a spike in rates of child malnutrition, the falling ratio of women to men, and dismal scores in youth education. Job creation remained stagnant. But the biggest black spot for Modi remains a 2002 pogrom in which more than a thousand people were killed, most of them Muslims murdered by a Hindu mob. While Modi himself was absolved by the Supreme Court of any direct links to the violence, there are many who believe he didn't do enough to stop the rioting, perhaps because the perpetrators were appeasing bloodthirsty factions in the state's Hindu majority. Either way, Modi went on to win a second term in Gujarat and then rapidly ascended to power in New Delhi. Whenever I chat with Hindu Gujaratis about the 2002 riots I am always struck by how they say "These things happen" in a tone approaching complicity. Gujarat, in my mind, has all too easily brushed 2002 under the carpet, a horror dismissed by a public jury obsessed with growth rates. (To be fair, independent India has similarly brushed other atrocities under the carpet, such as the 1984 Delhi riots against Sikhs, and the mass 1970s campaign of forced sterilizations, among many others.)

When I arrived at Ahmedabad airport, streamers advertising Vibrant Gujarat festooned the walls. There were dozens of colorful kites dangling from the ceilings to showcase an upcoming festival.

I had arranged for a car and driver to take me to Suraj. We were to to be joined by a local journalist I had contracted to spend the day with me, for I knew there would be some in the village who spoke only Gujarati.

We drove for nearly three hours to the district of Mehsana, where we picked up Pankaj Yadav. Pankaj worked as a local fixer, cameraman, and journalist all rolled into one. There were rarely any stories on his patch, he told me mournfully. He remembered the smartphone ban as news that had generated interest. Pankaj had a kind, eager face. He was happy to have some work and seemed thrilled to be partnering with a journalist from New Delhi. He had arranged for me to meet the current sarpanch of Suraj, a man by the name of Ramaji Thakore. (Devshi Vankar, the former sarpanch, continued to decline a meeting.)

We continued toward Suraj for another hour. Pankaj was an engaging—and amusing—travel companion, proudly pointing out which crops were growing, and how he estimated the newly constructed roads were bigger and wider than those in any other Indian state. Along the journey, clusters of solar panels caught my eye. It turned out they were scattered all over the area, powering the canals that diverted water from the nearby rivers. The fields adjoining these canals seemed particularly lush, growing wheat, cotton, and castor.

We had been traveling for several hours now so decided to stop for a chai before we entered the village. Jagat Singh Solangi not only ran the local tea stall but was also a trustee of a large temple nearby. He was an important figure in the area.

Solangi invited me to join a semicircle of his friends who were sitting on stools and drinking tea. It was midday, and the winter sun was overhead. A giant tractor was working just a few feet away from us, relaying a part of the road. Thick, tarry smoke wafted our way. Solangi and his friends didn't seem bothered; Pankaj and I tried not to think about the damage to our lungs.

A few sips of chai and some pleasantries later, I revealed to Solangi that I was a journalist and was here to investigate a story about a village that banned phones for girls.

"Bah! Media people," Solangi grunted, his voice getting louder as he wagged a finger at me. I realized he was putting on an act for his chums. "You people always show how women aren't respected. How they're being assaulted, raped. Can't you do any good stories?"

His friends laughed.

"Like cricket?" I suggested.

"Look, it's clear. Women shouldn't have smartphones. They don't have the mental capacity," Solangi said, pausing to see how his audience was reacting. They were nodding in agreement. "Maybe, just maybe, they can have a keypad phone, to stay in touch with the men. That's fine. But they don't have full knowledge to deal with a smartphone. If you show them the wrong things they won't be able to deal with it."

Full knowledge? What did he mean? And wasn't that the point of the internet?

Solangi gave me an appraising look. "Look, you're a grown-up man, so I'll be frank with you. Let's say I'm watching TV. Something vulgar comes on. Then I need to pee. So I have to step over my sister, my daughter. Now, if their clothes are even slightly astray, what if I have vulgar thoughts?"

The men started laughing.

"The devil will rise in me!"

I didn't quite understand what this had to do with phones, but the men were howling with laughter. It was clear Solangi was their local raconteur in chief.

"Just go see what these women are doing on their smartphones," he said with animation. "See the things they can access! What ideas will they get in their head? They're not equipped to deal. What if young women run away with men who have evil thoughts?"

While Solangi droned on, other men dropped by for chai, listened some, cackled, and moved on. By now the smoke from the tractor working the road had thickened to giant black clouds that almost enveloped us. I coughed and used that as an excuse to get out and return to the car. Solangi was hoping we could stay for lunch. A visitor from the capital was good for his image. I said I'd try to stop by on my way back, and escaped.

Dust swirled around Suraj. As our car drove toward Ramaji Thakore's home, the tires kicked up a swirl of loose dirt and sand. Through the haze, I could make out a few makeshift temples, scattered every couple of hundred meters, mostly one-roomed pink or yellow brick constructions. I was later told these shrines housed idols of local saints like Bapa Sitaram and the Hindu god Ram, hero of the epic *Ramayana*. They were competing with each other for influence and patronage, yet they were pitiable sites, run-down and forlorn.

Ramaji Thakore had been the sarpanch of the village on and off for the last fifteen years. His father had played the role for nearly four decades before him, running the affairs of the small village. Thakore had replaced Devshi Vankar a few months ago, perhaps as a result of the unwanted national publicity from the smartphone decree.

"We don't have elections," explained a young man outside Thakore's home. "We have selections." He grinned as he said this, speaking in Hindi but using English to emphasize "election" and "selection." He turned out to be Thakore's son, destined to be selected for years to come.

Thakore himself cut an impressive figure. He was well over six feet tall and wore a yellow *melia*, a large Gujarati turban with a splash of bold red embroidery. He had a *vari* in each ear, a pointed silver ear jewel shaped like a hook. Thakore was a *kshatriya*, a descendant of a warrior class that placed high in the hierarchy of the Indian caste system.

Life had aged Thakore. I learned that he was fifty-four, but to my eyes he looked in his eighties. He walked slowly and in a stoop; his senses seemed dulled and slowed. Gujarat was a "dry state," with alcohol officially banned, but I got the distinct sense that this had not stopped Thakore.

We walked across the village to the local dairy, which also served as an office for the village council. I had hoped that Pankaj and I could isolate the sarpanch for a short while and chat. Typically, we were not to be left alone. Thakore's sons and nephews were there as well, and they crowded us into a tiny room. We sat on wooden benches. The walls were finished in a chalky lime yellow; when we touched them, powder coated our clothes and bodies. On one of the walls was a calendar with a giant image of a beatific young Krishna, playing the flute as his companion Radha coyly nestles her head into his neck. In Hindu mythology, Krishna is the eighth avatar of the Hindu god Vishnu, sandwiched between other avatars like the god Ram and, later, the Buddha.

Thakore's sons were menacing, thuggish. For many of my first few questions, about the village, their lives, what crops they grew, they would answer before their father could. Their answers betrayed annoyance at our presence.

Thakore spoke, finally, when I asked him directly about the ban on phones.

"No such thing," he said, as Pankaj translated his Gujarati to Hindi for me.

"But the news? All the reports?"

Thakore stared at me for a moment, assessing.

He shook his head no.

Thakore's sons went on offense again. "Why are you here? Why are you spreading lies?" They were clearly trying to protect their father. "See any of the women here, there are no problems in this village. We are 100 percent successful!"

I motioned to Pankaj for help. Mercifully, he got the message and tried to distract the sons, walking them outside to ask them about something. They acquiesced.

If I thought I would now be alone with Thakore, however, I was wrong. Two of his friends wandered in. They were dressed similarly to Thakore: khaki kurta shirts, large turbans on their heads, and heavy silver jewelry on their ears and belts. Together, they were intimidating.

This time, language was a problem. Thakore could understand my Hindi, but when he spoke I could understand little of his Gujarati.

Finally, as I asked the same questions in different ways, one of his friends intervened. He could speak Hindi and seemed to feel sorry for me.

"Look, yes, we have a ban on phones for girls. What's the problem? It's true. You people warp everything and you don't understand. The reality is, several of our girls had run away with people they were talking to on the phone."

Thakore understood what his friend said and motioned him to stop. The friend was instantly silenced.

The truth, once out, lingered in the air.

How many girls ran away? What a tragedy! Why, what happened? I persevered.

Once again, the friend: "Four or five. We spend all our lives working hard, feeding the families. Can you imagine what it's like when one of our children runs away? A daughter is an *amaanat*, an asset. We have to protect our assets. This is why we take away their phones. Everyone agrees. It's unanimous here."

But why just the girls? What have they done wrong? Why not the boys?

The friend: "It's not safe. Girls don't have the knowledge to deal with situations. Men lure them away. They're immature. They don't have sense!"

Again the issue of "knowledge." What about the good things that came from girls having smartphones—education apps, school prep, reading about the world, videos and movies?

This time, the second friend snorted, jumping into the conversation. "Pah! They shouldn't even go to school. That's where all the problems begin. It corrupts their minds!"

Thakore cracked a grin—one of the only expressions I'd seen him make. It seemed to open things up.

Some of the young girls in the village had been rebelling against the strictures of village life. They didn't want to be forced into marriages as young teens. They wanted to go to college. Some of them wanted to teach or become doctors. They had heard stories of women in India's cities, and how some had become successful businesswomen, sports stars, actresses. Why couldn't they try too? For men like Thakore, this was a ridiculous proposition, threatening the core platforms of local life. Some of the girls had run away, just weeks before their arranged marriages were to take place, unions that they never wanted; the failed setups had created immense embarrassment and much pointing of fingers.

"See, the phone itself is the problem," said the first friend, as Thakore now signaled his permission. "It was fine as long as they used it only to keep in touch with us. But now they do all these different things on it. WhatsApp, phone calls, Facebook. Why do they need to talk to strangers? Have you seen them on it, totally immersed, ignoring everything else? It's ruining them. They're losing their values."

The friend continued. "You know the god Ram? When his wife, Sita, was kidnapped, he made her sit in a fire to prove her purity, to make sure nothing untoward had happened to her. Such was his control. Such was her dedication! Can you imagine that today? No, we're losing everything!"

(The story of the *Ramayana* is compulsory reading for children in India. In the epic, King Ram's wife, Sita, is kidnapped by the demon king Ravana. The battle to defeat Ravana—and win back Sita—is the main plot of the epic. But right at the end of the story, when Sita is rescued, Ram questions her virtue. Sita, in a bid to prove her innocence, says she will sit in a raging fire to prove that she remained "pure." She is untouched by the fire.)

The friend continued: "If young girls are on phones all the time, playing music, dancing...what will people think? Who will marry them? Who will want them if they act like whores?"

It appeared there was only one good thing to emerge from women having phones. "We can call our wives to bring us food while we're working in the fields. That is permissible."

Given that the move to ban phones for girls was "unanimous," I asked if I could speak to any of the village women.

Silence.

What about boys? Why should they get to use phones?

Silence.

Why did they think girls tried to run away?

Thakore fumbled with something in his belt.

I took this to be my cue to leave. I asked if I could stop by the local school. Thakore seemed amenable to the notion, so I was escorted across the street.

The Vidyalakshmi Bhavan school (*vidya*, "knowledge"; Lakshmi, goddess of wealth; *bhavan*, "home") was a collection of single-story rooms around a courtyard. It was an institution that practiced the midday-meal scheme, a government program that offers schoolchildren a free hot lunch. More than 120 million Indian students get these free meals; they provide an incentive for poor families to send their children to the schools.

The school principal was a middle-aged, avuncular man. In Thakore's presence, he seemed meek. No, there is no ban on phones in the village, he said, lying blatantly. No, no one ever runs away here! Yes, many of the girls who graduate from the school go on to become doctors and lawyers and engineers. Thakore listened in, glassy eyed, seemingly forgetting the conversation we had just had, happy to return to the world of make-believe.

A while later, perhaps as a reward for sitting and hearing paeans to the village students' accomplishments, I finally got a chance to speak to the only woman I would meet that entire day. Pankaj, thankfully, managed to distract the others for a half hour.

The woman—I won't name her here—had been teaching at schools in the area for a decade, commuting daily from her home in another village. Her head was covered with a part of her saree, her hair tightly wrapped up almost as if by a bandana. As she spoke, with great empathy, my surroundings finally came alive, the haze of lies cleared away: alcoholism, men beating their wives evening after evening, marital rape, rampaging upper-caste men, women doing all the work in the fields and at home and everywhere else, girls running away because they don't want the lives of slaves. And yes, the ban on smartphones. "Can you imagine how I feel, coming here every day? I knew each and every one of the girls that ran away. I taught them! They didn't run away because they had smartphones. They ran away to escape the lives they were destined to have in this village. Anything was better than what they would have here," she said, her voice lowered, making sure she wasn't heard. "We just have to keep fighting. All I can do is keep trying to infuse these young children with the right thoughts. Especially the boys. They are the future. They're the ones who can change this system of misogyny. What else is there to live for?"

India's women represent the world's largest untapped market for the internet. But reaching them, Google knows, is a complex challenge. "There are millions of women across the country who have to fight social and cultural norms every day," acknowledged Sapna Chadha, Google's regional marketing chief, at a company event in New Delhi in December 2017. Chadha was describing the success of the Internet Saathi program. "What's heartening to note is that these restrictions have not diminished their passion to learn new skills and pursue new opportunities. And in their pursuit they see the internet as a force to learn and to create a better future for their families."

Clichéd though it may be, the notion of the internet as a symbol of hope is wondrously alive in Phoolwati's village in Rajasthan.

Appearances can be deceptive: entering Nangli Jamawat is like being confronted with an older, forgotten India. The village begins as a muddy offshoot of a paved road. The track is narrow, unsure. Mounds of cow dung dot either side. The deposits of dung are surprisingly ornate: pies of fresh manure are patted with the crisscross of fingerprints and then neatly arranged in circles, coiling up in a cylindrical shape to rise as high as four feet. The painstaking effort to dry the cakes of dung showcases not only their value as disinfectant, fertilizer, and fuel but also a lack of modern alternatives.

The cow, as provider of dung and milk, is revered here; it is a symbol of prosperity and health. I remember once being struck by the sight of a lumbering, bejeweled animal, her head covered in a colorful, sequined shawl. Like a young bride, the cow sported on her forehead a red mark made from the paste of *kumkum*, a powder worn by Hindus after prayer.

In the winter months, mustard flowers carpet the fields of Nangli Jamawat in a vibrant yellow. Set against this backdrop, the women stand out in their pink and red sarees as they carry their brass pots to and fro. When they notice my gaze they cover their heads with a length of cloth, peering through what I imagine to be a rose-tinted veil. The men are altogether less remarkable, dragging their plows, with loose white dhotis wrapped around their waists and legs, cotton kurtas flowing on top. The men and women alike have faces like worn leather.

Farther along the trail, small thatched huts start to appear like dry sores on the land. A few are made of brick and cement, but they are modest, single storied, and unpainted. The clusters of brick homes are segregated from the huts, a reminder of how different castes mind their distances.

Nangli Jamawat is tucked away in an area known as Haldina, about fifty kilometers from the town of Alwar in Rajasthan. To see Phoolwati, I visited the village on and off over a period of about a year. If I left before dawn in New Delhi, I would make the drive in four hours. It often seemed to me those four hours were like a journey into history, forty or perhaps even four hundred years ago, an era without the comforts of modernity, electricity, and running water—let alone washing machines, microwaves, or TVs. Where Indian cities are densely populated and chaotic, the village is sprawling and sleepy; if city dwellers seem driven by ambition and accumulation, the people I met in the village seemed passive and fatalistic. Some of these perceptions revealed my own biases, since much of my lived experience in India has been in the big cities. And yet I knew villages like Nangli Jamawat were not part of an imagined bygone era; they are a reality for untold millions of families across the country, their version of India simultaneously coexisting with the brash, shining New Indias that feature in TV commercials and the movies.[3]

Phoolwati's home is located a few hundred yards into Nangli Jamawat, revealed by a large courtyard. Phoolwati slept with her husband and two young children in the room with wooden bed frames and colorful mats.

In the other, her mother-in-law often sat on a jute *charpoy*. The walls were unadorned. The cement floor was cool and uneven. Outside, adjoining the fields, was a small open-air enclosure that served as a kitchen. Farther away, in the middle of a courtyard, a makeshift hut housed a toilet shared by several families living nearby.

The more I saw of Phoolwati, the more I admired her fierce air of resilience. The skin on her face was taut and stretched, so much so that when she grinned or grimaced, it seemed to be holding her back, restraining her emotions. She wore what was standard in villages like hers, red patterned sarees expertly draped all over, loose ends wielded like magic to cover bare skin.

Phoolwati's days began like those of many other young wives in rural India. She woke at dawn. If she slept in, she missed her turn at the communal toilet and was forced to walk out into the fields to relieve herself. Phoolwati's first morning task was to bring water from the well, which was some way off; none of the village homes have running water. Then she fed and milked the cows. She would make a small fire from dung cakes to prepare her first batch of sugary tea. Phoolwati would save some milk for the children and set aside the rest to make yogurt. By the time she had kneaded dough and made rotis for breakfast, she had already been hard at work for four hours. After she served the family, she would eat last.

"This is what we do," Phoolwati said, speaking matter-of-factly for the women in her village. She might as well have spoken for tens of millions of India's women. "This is our routine."

One day, I followed Phoolwati on her rounds as an internet saathi. She cut a striking figure on her bicycle, saree bunched up to free her legs. She has a small, wiry frame and seemed perched uncomfortably atop her saddle, but her expression was set with determination. It is not common to see women travel on bicycles in Indian cities, let alone in villages, partly because it is viewed as immodest but mainly because it could be unsafe, with predatory men in abundance.

When I asked her about this, Phoolwati brushed off any concerns. "I have work, so I have to work," she said. "Who has time to worry about what people think?"

I asked her if she had heard some of the stories of sexual assault and rape that had dominated Indian media coverage. Phoolwati did not follow the news, but she had heard of the most infamous of the cases. In December 2012, twenty-three-year-old Jyoti Singh went to watch a movie at a Delhi multiplex. Singh was a call center worker by night and was studying for a degree in physiotherapy by day. Even though she came from a poor family—her father made less than $100 per month working at the airport—she represented the modern ideal of social mobility, destined to prosper in the city. After the movie, Singh and her male companion got on a bus to go home. But six men, including the bus driver, were already on the bus that day, high on drugs, cruising the streets, spoiling for a fight. After beating up Singh's friend, they took turns raping Singh. Her injuries were so severe that her internal organs were damaged beyond repair. Then they abandoned her to die. Singh fought for a few days in the hospital before succumbing to her injuries. The media, in days of print and TV coverage, dubbed her "Nirbhaya"—the one without fear.

"As an internet saathi, I spend most of my time with the village women. But if a man even dares to try something with me I'll give it right back to him!" Phoolwati declared, motioning with her fist. I couldn't help but grin at her response. But Phoolwati then got serious. "I worry more about my daughter, Pari." Pari's name means "angel." She was seven. I had met her before at Phoolwati's.

Phoolwati was right to worry. In 2015, 34,651 cases of rape were reported—about one every fifteen minutes. That same year, the last for which the National Crime Records Bureau made data available, 130,195 cases of sexual offense were reported. There were 327,394 cases of "crimes against women." For all these crimes, conviction rates hovered around a disappointing one in three.[4]

"Can't the smartphone keep women safe?" Phoolwati asked me. At the time, we dismissed the suggestion, but when I checked later, I learned that there were indeed apps that could make women at least a little safer in India. A start-up called Nirbhaya: Be Fearless created a set of apps to help users alert friends and family if they feel unsafe. On one of their apps, hitting the panic button ensures GPS coordinates are sent to a preselected group of people who can either come to one's help or alert the police.

Another smartphone app, Safetipin, lists safety scores of different parts of Delhi on parameters such as lighting, openness, transportation, and security. Safetipin had partnered with Uber in India's cities, but it was not yet available for a place as remote as Nangli Jamawat.

Spending time in rural Rajasthan and visiting places like Suraj in Gujarat highlighted to me the many challenges India's women face every day. While some struggle to get their hands on a simple phone, others face restrictions from their families, with little agency to change their circumstances. Various international rankings reveal a horrific national tapestry. The World Economic Forum's Gender Gap Index places India 118th out of 144 countries for female literacy; Indian women are ranked 96th for enrollment in tertiary education and 136th in the world for economic participation—just ahead of Yemen, Saudi Arabia, Pakistan, and Syria.[5] In one remarkably grim statistic, India ranks 141st for sex ratios at birth. For decades, Indian families have aborted female fetuses to prioritize having male children. Despite government clampdowns, the practice continues. According to the government's 2011 census, there were 943 women for every 1,000 men.

I asked Phoolwati if she found any of the numbers surprising. She didn't. "The status quo is perfect for men. Anything that gives us freedom is a threat. People in our village also sometimes question my work as an internet saathi. But since my husband is OK with it no one dares to talk."

Phoolwati's husband, a jolly-looking man, would often be around when I came to visit. Satish is a farmer by trade but he also teaches mathematics at a school in a nearby village. He looks different from the other men in his family. Satish wears an English-style shirt, not a kurta, and trousers, not a dhoti. These are small but telling acts of noncompliance. When he looks at Phoolwati, even fleetingly, it reveals someone in love, someone proud and happy. He told me once that their daughter, Pari, was first in her class, that she's smarter than all the boys. He describes his wife as a change agent. He worries for her safety when she travels around the region but boasts that his wife is a fighter. Satish is strong enough to watch his wife work and prosper, going out into the world on her own.

Phoolwati is a success story, but without Satish, her path would be a very different journey. And the conditions would be different elsewhere. Phoolwati in Suraj, Gujarat, wouldn't be the determined Phoolwati I got to know; she would be cowed into submission. She would have tried to run away.

"The biggest sign of how the internet is changing India is women's empowerment," said Osama Manzar, the founder of India's Digital Empowerment Foundation, which has more than 150 centers across rural parts of the country. "They are challenging India's patriarchy. As soon as a woman has a mobile phone in her hand, she has more confidence. Many men want to quash that confidence."

Manzar is often seen at tech conferences in India wearing his signature patterned turbans. He, like many others invested in expanding the internet in India, says rural areas are a forgotten market—despite the fact that it is where the need is greatest.

"Imagine if you need to photocopy some papers in the village. Where do you go? If you want to watch a movie? Pay a bill? Where do you go? You have no options. Poor and rural Indians spend a lot of time traveling long distances on these simple things. They lose entire days of earning potential. But with the internet, things can get easier."

Women could be the biggest beneficiaries, in large part because India's internet boom has left them behind.

"In India, as soon as you are connected on the internet you leapfrog from being illiterate to educated," said Manzar. "Yesterday they were illiterate; now they are consuming and at par with educated people. If you and I are both connected, you are not going to ask for my caste. The internet in India makes people equal."

When I put Manzar's take to Phoolwati, she agreed. "Look, I can't speak for the whole country, but I can tell you what I see in the villages around Alwar. Women are left behind. We just want to have a chance," she said. "And with the internet, we finally have some straws to clutch on to."

Phoolwati hopped off her bicycle as she neared her home. We walked the last fifty meters to her house and the shed where she stored her bike. Phoolwati-the-Internet-Saathi seemed to become a different Phoolwati once she was home. Her head receded into her shoulders, tilted down in submission. Her walk changed from a bold stride to more of a quick shuffle. She seemed to shrink. Here, at home, she was wife-mother-sister-daughter-in-law. She shrouded her face with her saree.

Once inside, we continued to chat. As she often did, Phoolwati listed the things she had learned from the internet. There were recipes for dosas, the crepelike South Indian dish served with coconut chutney; she had studied which seeds her husband should plant in the summers, and she had figured out how to send pictures to people on WhatsApp. I suspected at least part of her list was from her training, regurgitated from Google's instructions for these sorts of things. Even so, Phoolwati's enthusiasm about the internet was unmistakable.

"This is my identity now," she said. "People in all the villages nearby know me as the 'internet lady.' I am showing other women what they are missing out on. I am showing them things they would never dream of seeing and hearing."

One of her students, Phoolwati said, was a widow who was unaware of the government benefits available to women who had been bereaved. Another didn't know there was a grant for building toilets. "We got 12,000 rupees from the government too, to help build what we have," she said, pointing outside. "I learned about this by asking Google. Otherwise, who will tell us? How will we know?"

But is the internet truly fixing her many problems? I asked, somewhat uncharitably. Is it making life easier? Is it changing anything for her and other women in the villages around her?

"Change comes slowly," she replied, lowering her eyes as she spoke. I suddenly became conscious that we weren't alone. We were chatting, one-on-one, sitting cross-legged on the floor of her communal space, but surrounded now by a gaggle of extended family, including her sister, sister-in-law, mother-in-law, and several village children. This tended to happen every time I came to see Phoolwati. I would ask for some time alone, she would agree, and then her mother-in-law would plant herself beside us, inviting others nearby to join in. After a couple of visits, I assumed this to be normal village behavior. No one spoke in private. They would hardly talk or make eye contact, especially the mother-in-law, who kept her head covered at all times, peering at me through her veil. The other women giggled at how I drank tea, not slurping noisily the way the men in their family did. Invariably, one of the children would borrow Phoolwati's phone and take pictures of us.

"Flesh-a-lilskee, flesh-a-lilskee!" sang one girl as her friend took a selfie with her lips in a comical pout. I had no idea what this meant. Some weeks later, while I was watching a cricket match on TV, an ad for the Chinese smartphone company Gionee came on. The commercial showed the Bollywood actress Alia Bhatt in a pink halter top taking selfies of herself while a voice sang, "Flash a little skin, flash a little skin ..." The village girls must have seen the ad on a smartphone, for Phoolwati's home didn't have a TV. They didn't speak English, so they probably didn't understand what they were singing.

When Phoolwati was not out working, her internet saathi tools were put to good use at home by the extended family. Her nieces—ten-year-old Manju and eight-year-old Sanju—would take one of Phoolwati's tablets and ask Google how to make samosas and embroider designs on their clothes. YouTube seemed to have videos for everything. Pari was using it to learn basic multiplication. And Phoolwati's son, Prashant, was only three years old but had figured out how to watch cartoons on his own. (YouTube pioneered "offline" videos in India as an antidote to slow download speeds. People set videos to download and watch them later, over and over again. This feature is especially useful for children with cartoons or adults with devotional videos.)

The homes in the village receive only six hours of electricity a day. When the lights flickered, there was usually a rush to charge Phoolwati's devices—the two tablets, the two phones, and the power packs.

I could only marvel at how many things the smartphone seemed to represent to Phoolwati's family: it was their TV, their camera, their music player. It was the home's centerpiece. It afforded Manju and Sanju some little privacy if they wanted to search for something without everyone knowing about it. And the gatekeeper to this world of discoveries was not a male village elder but a woman.

I asked Phoolwati's family what the smartphone represented for them.

Phoolwati's older sister Santoshi Devi spoke up now. "Hope," she said simply. Santoshi Devi could only read and write a little, far less well than Phoolwati. "Women here can't do anything on their own. We weren't allowed to study when we were young. We didn't choose who we married, we just said yes. We have no power. But maybe phones can help the next generation, somehow."

"The internet is the greatest invention of our lifetimes," added Phoolwati. "I'm doing this for the future. My daughter, Pari, will grow up in a different India—not thanks to the people or the culture but thanks to the smartphone. For poor people like us, this is a revolution. It is our independence. I'm the first woman in this village to work outside the home or the fields," she said. "Now every young girl in Alwar wants to step out and be an internet saathi like me."

2

Hello Macaulay

Fixing Education

"MY BUSINESS IS SIMPLE," said Abdul Wahid. "I . . . I . . . P."

He paused. I waited.

"Is . . . It . . . Possible."

As he said the words, Abdul Wahid drew in the air a grand billboard for his tutoring company. He paused once again for effect.

"Is what possible?" I asked, finally.

"Anything. Everything." And then he added: "It is a cun-sept."

Abdul Wahid was very fond of concepts.

Abdul—as he insisted I call him—was the son of a small *dhaba* owner in Kolkata. Ambitious, he didn't want to run a loss-making, hole-in-the-wall restaurant like his father, dishing out dal-roti-sabzi to customers who were rude and never tipped. He didn't want to do the same things every single day. Abdul wanted more. He wanted to be his own man. An entrepreneur. He had dreams that went beyond his father's street. With the help of his smartphone, Abdul had turned himself into an English-speaking teacher-CEO. And now, with his tutoring company IIP, he wanted to transform education in India.

Still, it was clear to me that the past clung to Abdul. He was only twenty-five, but he dressed like a much older man, in a crisply starched white shirt and pleated dark khakis. His eyes were almost comically enlarged by his old-fashioned rimless spectacles. When he moved, he gave off a faint aroma of cardamom; he carried pods in his trouser pocket, just as his father and his grandfather before him.

Abdul had a vision for tutoring in India. "I want to make education like a gaming platform. Indian teachers think in 2-D. I want to make it like a planetarium experience. I want kids to *see* and *feel* projectile motion. Is it possible? Yes, it is."

Money was scarce in Abdul's childhood. Back then, his dreams were dull, black-and-white, 2-D, standard definition 4-by-3. His father sent him away to a boarding school in a town in neighboring Bihar. Abdul found the system of rote learning a bore; his teachers were disinterested, playing truant more than he did. He still rubs his left cheek when he remembers the "tight slaps" he would receive for not reciting the twelve-times table all the way to the end.

One moment in his childhood was defining. Abdul was fourteen. He and his classmates were woken every morning at 6:00 a.m. by the school warden. The students, shivering in their underpants, would line up in a courtyard armed with steel buckets of cold water to wash themselves, mug by mug, bar of soap ready to vigorously rub away the shivers. Abdul was terrified of the chilly water. He began skipping days between bucket baths. His friends made fun of him. Abdul wouldn't bathe. They called him smelly. No one sat next to him in class.

Eventually, something had to give. Abdul asked himself if he could learn to conquer his fear of cold water. *Is... It... Possible?* Yes or no? A mug of freezing water, no matter how searing, fell within Abdul's binary realm of the mathematically possible. Logic afforded him no excuse. It was an epiphany, like a bulb being switched on, from zero to one. Abdul began bathing every day.

"Successful people always say 'nothing is impossible.' Bogus! Many things are impossible. But there are also things that are possible. We need clarity on those things."

"That's the power of IIP," he said of his company. "Possible or impossible? Answer the question, and you move closer to your dreams."

Abdul was not just a teacher; he was the motivator-dreamer-inspirer-in-chief as well as founder-CEO-owner of IIP, a "coaching center" an hour's drive from Jaipur, the capital of Rajasthan.

"I always ask my students every morning, *Is It Possible*? Can you wake up if you set an alarm for 4:00 a.m.? Can you give up the extra ten minutes of sleep? Can you do six hours of homework? Possible or no?

Haan ki na? Yes or no? *Is...It...Possible?* That is the concept," announced Abdul, jumping fluidly back and forth between Hindi and English. He speaks mostly in Hindi but uses English when he needs to call on a technical word or a cliché.

"Body is slave, brain is master!" he exclaimed, punctuating the air with his index finger.

Coaching centers like Abdul's IIP are part of an enormous parallel education industry in India. Every year, two hundred thousand students enroll in these institutes to learn how to ace admission tests for India's medical and engineering schools. The tutoring is intense, often lasting more than eight hours a day. The lessons are designed as a "hack" for the entrance tests: trick questions are demystified, shortcuts identified, problems rehearsed until they become muscle memory. Each tiny advantage is necessary because only a fraction of the applicants achieve success. At the famed Indian Institutes of Technology, or IITs, some five hundred thousand aspirants vie for ten thousand available spots. The 2 percent rate of admission makes these colleges more competitive—at least on paper—than Harvard, Stanford, and MIT. Many of the rejected applicants try again and again, year after year, their hopes fixed on the institutions that have churned out the CEOs of Google, Sun Microsystems, Hotmail, Infosys, and more.

Many of the successful candidates at the IITs and India's top medical schools come from coaching centers, which usually feature bluntly aspirational names: Brilliant, Vibrant, Acme, Mega, and, of course, IIP. Often students take a full year after high school to prepare for the university admissions tests. Many of these young dreamers—a potpourri of poor and rich, rural and urban, practicing a Babel of tongues—live in the cramped centers themselves or in hostels adjoining them. In Rajasthan, the hub for these coaching centers, entire towns have sprouted to support these institutes with twenty-four-hour convenience stores, shopping malls, *dhabas*, and hostels. At the centers, morning and night, teachers ply their students with diagrams and sample examinations, all designed to help them "crack" the entrance exams. The goal is not education; it's about *winning*. When students get in, their successes are trumpeted all over the country in newspapers and magazines, their pictures neatly arranged in rows and columns: "23 students in all-India top 1000! Highest rate of success guaranteed!" The winning mug shots often look the same: tightly coiled, bespectacled, scrunched-up faces, carrying the collective burdens of their families, teachers, towns.

The coaching institutes are the epicenters of India's age of aspiration; they represent the ultimate life hack, the fast track to start-up, IPO, millionaire. Students whisper among themselves tales of alumni who have gone all the way from small-town India to the top engineering or medical schools, and then onward to the United States, ending up with a mansion, complete with swimming pool, and an SUV. Teachers baldly describe "the dream package": a starting salary in high-six-figure dollars.

The centers exist because most students need more than their high schools are giving them. In fact, years before Indian students attend these coaching centers they have already gotten used to the idea of their regular teachers' inadequacies. According to the Indian government, nearly a third of all Indian students take private after-school tuition.

Abdul Wahid's IIP was an upstart, a tiny but rising star in a constellation of extracurricular options. In 2016, more than a dozen of his students were accepted to medical schools. He was aiming for a bigger "catch" this year.

"I used to be against coaching centers," he lamented. "But you have to understand the system to beat it. This is an important experience for me to enable my next dream—to change schooling in India."

When Abdul describes his coaching methods, he shifts into the soaring Hindi rhetoric he uses with his students.

"My kids are freshest in the morning. Like flowers."

I jotted down some notes. This pleased Abdul. He slowed down to help me out.

"As soon as they wake up, we give them six hours of nonstop instruction—biology, chemistry, problem solving. Then a quick lunch, and I put them to sleep for an hour."

"Why sleep?" I ask.

"When I wake them, it's a new morning. They're fresh again, in full bloom. That's why we can drill them again. Our teachings go right into their brains. I've created *two* mornings."

He smiled smugly at what I imagined to be a well-rehearsed line.

So, day and night, day and night, they just cram? I used a mix of colloquial Hindi-English to ask the question.

"Survival of the fittest," replied Abdul, reverting to English to recall a cliché.

Once, when I was interviewing him in his office at IIP, the parents of a prospective student dropped by for a chat. They had journeyed more than five hundred miles from Uttar Pradesh to drop off their son, and to pay a year's worth of coaching fees and room and board. The son had failed to get into medical school for three years running. Then they heard of another boy from their town, also someone who had been trying for years, who had now gotten into his dream university, a government medical school in Bangalore. How? He had spent a year with Abdul and learned the gospel of IIP.

Abdul allowed me to chat with the parents. "See: he's come all the way from Delhi to interview me!"

I asked the parents what attracted them to IIP.

"Discipline!" declared the father. "My son is useless. Duffer! Always distracted. How many more times can he fail? He needs to keep his nose buried in the books. This is the place to do it!"

The son looked at us and nodded, defeated. "I need to work harder," he mumbled. "I have to by-heart the whole syllabus." ("By-heart" is short for "learn by heart," or rote memorization.)

IIP was indeed the place to study, to by-heart, for there was little else one could possibly do. The coaching center consisted of two nondescript white buildings in a vast and abandoned housing project, placed off a highway some distance from the city of Jaipur—and not near Kota, where Rajasthan's more famous coaching centers were. One of the buildings, two stories, housed a collection of classrooms, along with Abdul's office. The other had six stories. Abdul was renting the first three floors as a hostel for his students. The walls were plastered white. Loose wires and cables draped the exterior. There were bunk beds in every room, home to six or so students each. There was no other furniture. Girls and boys were separated in different wings, but they all ate together in one kitchen mess, sitting on the floor as a cook dished out dal and roti. The austerity was almost a form of punishment, and punishment was what many of the students had signed up for.

I now asked the mother why they had traveled all the way from Uttar Pradesh to Rajasthan.

"You know the biggest thing? Wahid-Sir takes away their smartphones. Simply not allowed."

"Is that true?" I ask.

"Yes. Look at my son. All day long on YouTube and FB [Facebook], watching silly movies and downloads. How will he focus? Here, there is nothing else to do. No distractions. Wahid-Sir has created the perfect environment where all you can do is cram-cram-cram. Only then will my son get into medical school. We can't pay for expensive private schools. But if he studies hard and gets into a government medical school, then he will become a doctor and his life will be made."

As the mother teared up—rather theatrically—the father added: "He will go through pain. Hard work is a bitter pill. And what about us? A mother can't see her child go through pain, but what to do? This is necessary. All other options have failed."

Abdul was smiling through this exchange.

"It *is* possible," he pronounced. "He can make it."

And then: "Hard work beats talent . . . if talent doesn't work hard." This was one of Abdul's many preprepared English zingers. His audience was suitably impressed. I took my cue and kept taking notes.

I asked him why it was so important to take away the students' phones at the start of the semester.

"It's a time waster," he said of the smartphone. "You can say wow, so many good things on it, physics, chemistry, biology apps, *wow*!"

Abdul loved to use English words and phrases every now and then. "Wow" was a particular favorite.

"But it's a big distraction now. Some students spend all their days on WhatsApp. They have a fascination with the opposite sex. They download movies. These students didn't grow up with these things, so they are distracted," he says. "We have to make them focus!

"I didn't want to, but I was forced to ban the smartphone in my institute," said Abdul, before adding: "For now."

For now, because Abdul has big dreams. In the future, once he has a track record of years of success with students settled in the United States and top hospitals in India, he wants to build India's first smartphone-driven educational university, with campuses all over the country.

"Students will walk into the auditorium," he later described to me. "There will be smartphones on stands for each student. No FB, YouTube. Only educational apps. There will be an app to project video on the skies, teaching physics and mathematics in 3-D. Mind games. Vedic mathematics. The works. It is the future."

Abdul's eyes glistened with hope. "Is it possible? *Yes!* That is the concept."

February 2, 1835, was one of the most consequential days in Indian history. The events of that day shaped Indian education for the next two centuries—and provide a backdrop for why smartphone use is now so transformative.

A few months earlier, Thomas Babington Macaulay, a thirty-four-year-old British statesman and orator, had moved to India to serve as a member of the Supreme Council, which reported to the East India Company's Court of Directors. At the time, the East India Company controlled vast swathes of the lands we now know as India, Bangladesh, and Pakistan, from Dhaka in the east to Lahore in the northwest. On this day, Macaulay was to address the council amid a debate over funding the study of English—instead of Sanskrit and Persian.

Here is an excerpt of what has come to be known as "Macaulay's Minute":

> All parties seem to be agreed on one point, that the dialects commonly spoken among the natives of this part of India contain neither literary nor scientific information, and are moreover so poor and rude that, until they are enriched from some other quarter, it will not be easy to translate any valuable work into them. . . .
>
> I have no knowledge of either Sanskrit or Arabic. . . . It is, I believe, no exaggeration to say that all the historical information which has been collected from all the books written in the Sanskrit language is less valuable than what may be found in the most paltry abridgments used at preparatory schools in England.

Part of Macaulay's assessment was based on what he believed Indians wanted. "All the declamations in the world about the love and reverence of the natives for their sacred dialects," Macaulay declared, "will never, in the mind of any impartial person, outweigh this undisputed fact, that we cannot find in all our vast empire a single student who will let us teach him those dialects unless we will pay him."

The solution, then, was to teach Indians English. But given that it would be impossible to educate everyone, the British would create "a class of persons Indian in blood and colour, but English in tastes, in opinions, in morals, and in intellect." Macaulay was effectively arguing for the perpetuation of empire through language. The few, ruling the many, forever. As the British weekly *The Economist* put it in 2004, Macaulay was being a "supercilious English xenophobe."[1]

Macaulay carried the day. The empire began funding English classes across the country, aimed at training the select few, thus beginning India's tradition of English haves and have-nots. As Abdul put it to me in his inimitable way, Indians would "divide and rule" each other, the English speakers ruling the hoi polloi forever more.

The smartphone gave Abdul what Macaulay had envisioned for a small elite.

Hello English is an educational smartphone app—among the most popular in India—that teaches basic English through a series of tests and games. A prospective student only needs to know the alphabet and be fluent in at least one of twenty-two Indian languages. The app offers an array of word and sentence games; it latches on to your level of English fluency, targeting and then improving weaknesses.

Like many games, the app ranks its users. Of its 15 million subscribers worldwide, Abdul ranked number one in February of 2015. He was the poster boy of smartphone-based English learning.

"I played twelve hours a day, every day. Wherever I went I played Hello English—in traffic jams, in bank queues, in the bathroom," remembered Abdul. "My English was weak. But without English I can't impress clients in India."

And English, Abdul understood, was the great divider in Indian education. "We are a discriminating country. When we study in school as kids we learn our history and physics and biology in Hindi or Bengali or Telugu," he said, referring to the most-spoken Indian languages. "But when we apply for graduate schools, the tests are in English. The engineering and medical colleges teach you in English. Suddenly, you have to relearn all the technical terms and concepts in a foreign language. How is that possible?"

A majority of Abdul's students struggled with English; most came from what is known in India as "Hindi-medium" schools, where the language of instruction is, of course, Hindi. As a result, they learned only cursory English. Now that they needed to test their physics and biology in English, they were struggling.

On the other end of the spectrum are Macaulay's elite, the urban students of "English-medium" schools, where attendees learn their history and science in, naturally, English. Graduates from these schools often go on to the best colleges in India, almost all of which also instruct their students in English.

Speaking English fluently is the safest guarantee for white-collar work in modern India. Until recently, it was the gateway to the internet. It also allows Indians to communicate with their peers across the country's twenty-nine states. Students of Hindi-medium or any other non-English-language-medium schools are herded into a self-perpetuating lower caste, right from kindergarten.

Abdul was an outlier. He had broken through the divides of his childhood. Just a few years earlier his English was weak and faltering. Now he was impressing his students and their parents with his speech. He seemed almost Gatsby-esque. Instead of bootlegging, however, he would use the internet to reach his green light.

Abdul's IIP website, ConceptOfIIP.com, showcases a motto on its home page. It is a quote from the American writer Ta-Nehisi Coates, with the header "Redefining Possibilities."

> The pursuit of knowing was freedom to me, the right to declare your own curiosities and follow them through all manner of books. I was made for the library, not the classroom. The classroom was a jail of other people's interests. The library was open, unending, free. Slowly, I was discovering myself.

Fittingly, Abdul had not spent much time in classrooms. He had never liked school, never attended college. Abdul Wahid had not studied past twelfth grade. Of all the teachers at IIP, he was the only one without a real degree. Yet he presented himself so convincingly, so impressively, to droves of students and parents that they signed up with thousands of dollars in down payments for his annual course.

Abdul Wahid, by his own admission, owed everything he knew to the smartphone. When he graduated from high school in 2009, he took a couple of years off to prepare for the national medical school entrance tests. He ranked 2,177th in West Bengal's state examination, among the top 10 percent of test takers.

"I could have gotten into a good medical school. But I stopped. I was totally passionate to open I.I.P.," he said, emphasizing the words "totally passionate" in English. "But how could I just start like that? I needed first to learn many things."

Abdul didn't have a computer at home. He pooled together money to buy a smartphone and a data card. Then he dived into the biggest library of them all: the World Wide Web. "Whatever knowledge I have right now, 90 percent of it comes from YouTube and Wikipedia. That's how big a role it played in my life."

One of Abdul's heroes was Steve Jobs, the founder of Apple. "He had a vision. Stupendous! Wow!"

He also admired Sal Khan, the founder of the online-education-video company Khan Academy. "Sal Khan made me give up medical school. He taught me the power of the smartphone."

Every week, he would watch dozens of TED Talks, lingering especially over the ones that he found inspiring. He watched speeches by Indian motivators such as Shiv Khera, author of *You Can Win*. He watched Khan Academy tutoring videos and signed up for free online courses from Harvard and MIT, not to study the material but to master their methods. Abdul became a self-improvement junkie, rehearsing in front of a mirror lines he would use to inspire future students and their parents.

"You see how the parents respond to me, how they are impressed with me? I have no degrees. Parents are not inspired by degrees. They are inspired by passion.

"All my teachers have major degrees. But they don't instruct me, I instruct them. I have far better knowledge than my teachers. Knowledge matters, not degrees."

Surely, I pointed out, what I.I.P. was doing—forcing students to cram physics, chemistry, biology sixteen hours a day—wasn't imparting knowledge? Were his students learning about the world, about becoming well-rounded citizens?

"No," Abdul admitted. "You are right. This is the first step. I come from a poor family," he said, pointing to his ramshackle campus, a far cry from the richer institutes nearby in the town of Kota, where the likes of Acme, Brilliant, and Champion were paying their best instructors tens of thousands of dollars a year—a huge sum of money in India.

"First I must establish IIP. I need more students. I need more funds. Then I will reform the entire education system."

Abdul often used a stylus on his tablet to describe his ideas to me.

There are three types of kids, he explained once, as he drew on his digital blackboard.

"A is super intelligent. B is intelligent. C is average. D is below average," said Abdul. I didn't stop him to point out he had named four students, not three. He continued: "Each student gets the same one-year syllabus at the same time. A finishes in three months, B in four months, C in twelve, and D takes two years. What did A do for the rest of his year? Nothing! C has become a limiting agent. And D has failed. Why should A stop for B and C? Why should D fail? A day will come, Ravi, IIP will have its own national curriculum, and millions of students will sign up. Everything will be on a smartphone. If a student finishes a syllabus early, he will progress to the next level. It is like a game. That is the concept. The school won't decide the level; the student decides. That is what India is lacking. I have big investors ready to pay me for this idea."

Abdul's students seem to be fans of his ideas. On IIP's website, Faheem Akhtar called Abdul "awesome," adding that his program was "really unique" and helpful for "repeater students." Toshi Sharma wrote that IIP "gives us confidence that we can DO it." Sachin wrote an ode to Abdul: "all happens only bcz of u all sir... thanks for mking my dad's dream cm true."

"It is all because of the smartphone," says Abdul. "The smartphone gave me the internet. It gave me knowledge where I had none. It is the best concept in the world."

Most tech companies in India are housed in swanky high-rises. Hello English's Jaipur-based offices were comparatively discreet, nestled inside a sprawling old bungalow just two stories high. The building was a pale yellow, almost blurring into the blinding sun. A security guard pointed me toward the office entrance, up a narrow flight of stairs. I guessed that the owners lived in the bungalow's ground floor. The arrangement reminded me of many family-run Indian businesses, selling everything from linens to cricket bats.

The offices themselves were clean, efficient, modest. I was welcomed by the assistant who had set up my meeting. He took me to a small kitchenette and asked me to help myself to water and coffee. This was unlike most large Indian offices, where attendants would bring steaming, milky-sweet chai or coffee. Hello English seemed more like a Brooklyn basement start-up, without the exposed brick and the leather beanbags.

A few minutes later, I was joined by Pranshu Bhandari, cofounder of Culture Alley, the company that created Hello English. She ran it with her cofounder and husband, Nishant Patni. Unlike many other married Indian women, Pranshu had kept her maiden name.

I started with the obvious. I wanted to understand why two entrepreneurs from India decided to create a tutorial app for English, a service that was already well represented by companies like Rosetta Stone.

Pranshu replied with a story. In 2012 she was working as a consultant in Gurgaon, the satellite city near New Delhi that had attracted hundreds of international tech and management firms. Nishant, then her fiancé, was studying at the Kellogg School of Management in Chicago. As part of the program, he had to spend some time traveling through China. Nishant was struck by how completely unable he was to communicate with the Chinese: his English, Hindi, and Marwari were useless.

Why not create a Chinese language app? It was the world's hottest economy, after all. Pranshu would join him, and they could start a company together. They began their research and discovered how wrong they were. "The audience wanting to learn Spanish or Chinese as a second language is tiny. Those are exotic needs," she said. "English is a *roti-kapda-makaan* need." She meant that it covered basic necessities: *roti*, bread; *kapda*, clothes; *makaan*, house.

"Me learning Spanish and failing at it won't change my life," added Pranshu. "But an Indian trying to learn English and failing could impact his job prospects. Even for a housewife or a mom, it impacts her confidence."

Hello English had a ready-made audience: Indians who were discovering the internet on smartphones. These were Indians who had never used computers or iPods; Rosetta Stone was out of the question for them. They could not afford private tuition, and they had very little spare time. A free, easy-to-use app made perfect sense. It could be used on a bus, waiting in a queue, anywhere. It was a market that was guaranteed to grow and grow.

"The first question potential users asked us was 'Will I need a data SIM?'" said Pranshu. Internet data is set to get cheaper in India, but it remains expensive enough for poorer Indians to ration their consumption. "So we coded it in such a way that the app could work offline—that's what people want in India."

I was familiar with this offline availability from speaking with several other business leaders. YouTube, as noted earlier, has a feature where users can save videos for repeated offline viewing. A local CEO told me an anecdote that explained why this practice was so popular. A year back, he had opened a small office of about forty people to run a logistics business. Naturally, his office had free Wi-Fi for his employees. When he checked to look at how the internet was being used, he noted that there were big spikes in usage at 10:00 a.m. and then again at 6:00 p.m. He asked around to find out why. It turned out many of his workers had no internet at home and were unwilling to pay much money for data on their smartphones. So they would download all their mail and software updates when they came in to work in the morning, on free Wi-Fi; then when it was time to go home, they would download music videos to watch offline on their train commutes home. This was how millions of Indians were experiencing the internet.

"It's not just being able to use an app offline," said Pranshu. "Apps for Indian users also need to take up very little memory space. Most Indians still have very basic phones, and they ration space for only the most special apps."

Hello English is indeed a pared-down app. When I started using it I discovered that the app wasn't available on iPhones. More than 80 percent of Indians use Android, and that's the market app makers target. "If you want to understand how real Indians use phones, buy a $100 Micromax phone," chided Pranshu, tut-tutting at my iPhone. "Only then will you know how they really experience the internet!"

Indians have adapted to technology in a way that is thrifty and innovative. In a sense, that's how they have been learning languages too.

Most Indians have a peculiar way of speaking English. They think in their mother tongue, say, Hindi, and then translate it to English. The result is that you will often find Indians asking you "What is your good name?" The "good" is a literal translation of the same question in classical Hindi. Or you could find a taxi driver introducing himself as "myself Ramesh"—again, a literal, word-by-word translation of a colloquial vernacular introduction.

While Hello English has many different features, its most commonly used are the ones that correct these mistranslations. One lesson featured a game that showed a cup of chai emptying out as I was picking the correct answers. A Hindi sentence would pop up in a bubble, steaming above the tea as I was asked to pick the correct English translation. "What is your name?" or "What is yours name?" "My name is Neha" or "Neha my name." And so on. The more correct answers, the more gold coins one wins. A few games in and I found myself still outside the top 10 million users.

"As long as you read any Indian language, you can start using Hello English and progress upwards," Pranshu told me in an encouraging tone. "Indians are very practical. They don't want to read Dickens. They want to be able to communicate with each other."

At this point, Pranshu's husband, Nishant, joined us. He had just wrapped up a meeting with his coding team. Nishant had the calm demeanor of an older Indian businessman, though he had just turned thirty-one.

Both Nishant and Pranshu are Marwaris, an entrepreneurial group of Rajasthani Hindus who spread across India in the last century. Marwaris today run several of India's biggest family businesses. (I'm Marwari too—but as a journalist, a clear outlier.)

One conversation I had with a Marwari businessman drove home for me how the community viewed business. I was attending a cousin's wedding in Kolkata, having traveled from London, where I was based at the time as a TV producer. Visitors to India from abroad are often accorded lavish attention, and this was such an occasion. As the night wore on, my hosts introduced me to several other special guests. One of them—clearly a power-broker Marwari businessman—wasted no time sizing me up. How much did I make? He was unimpressed. Why do you do this work? I like it, I explained. This dissatisfied him even more. And then he said: "Look, you can be as successful as you want.

You can rise. But ultimately, there is only word to describe what you do: *service*." He said that last word with an exaggerated disgust, as if to imply my reporting to a manager was something dirty. He explained: "I run my own business. I report to no one. *That's* the life."

I put that exchange to Nishant. Did his heritage help him start a company?

"See, the concept of starting something from scratch is familiar to our community," he said. "The risk appetite of our families is pretty high," he added, pointing to Pranshu as well. "Maybe it helped us start Hello English. But most of all, we did this because it was our passion. And the demand for a service like this was so clear to us. People here have always wanted to learn English, but they often didn't have a way to do it. Now with the smartphone, they can; it's democratizing everything in India. That's why we started this app. This is where we have the greatest chance to impact the greatest number of people."

Numbers always let India down. The country now has more than 1.33 billion people. As noted previously, only 73 percent of Indians over age seven are functionally literate. Of the 900 million Indians with some basic literacy,[2] an estimated 200 million can read or write rudimentary English. The perceived notion that Indians—compared with Chinese or Indonesians or Koreans—are largely fluent in English is heavily biased toward an urban, upper-middle-class sample pool. The reality is that most Indians speak no English at all. White-collar jobs in offices, tech companies, or call centers are therefore off-limits.

What about schools? Here too, the data is damning. According to the World Economic Forum's 2017 Global Competitiveness Index, out of 137 countries surveyed, India ranked 49th in the world for the quality of its primary education.[3] It fared better in the quality of its higher education, at 26th in the world, but very few Indians could avail themselves of this opportunity: India ranked 97th in the world for enrollment. Only 18 percent of Indians make it to college. (By comparison, 29 percent of the world's population attends university. India is the biggest drag on the global figure.)

Several Indian CEOs have lamented that most of their country's graduates are almost unhirable, needing months of training and coaching before they can be of use. Devesh Kapur, a professor of political science at the University of Pennsylvania, has pointed out that between 2000 and 2015, India opened six new colleges every single day, including weekends. But the quality of the schools was pitiable. "We have the second or third largest country of people with college degrees in the world," he told the *Hindu* in a July 2017 interview.[4] "But everywhere, whether public or private institutions, we have a shortage of talent. You know that old poem? Water, water, everywhere, not a drop to drink. We have graduates, graduates, everywhere, but who do I hire?"

Higher education in India has severe problems, but the underlying problem begins much earlier for most Indians: at primary school.

The economist Karthik Muralidharan of the University of San Diego has emerged as a global expert on this issue, publishing dozens of highly regarded papers on teacher truancy and the quality of Indian education. In an influential paper summarizing research on Indian education, Muralidharan pointed out that India had made several vital improvements to its infrastructure.[5] Pupil-teacher ratios had fallen by 20 percent over a decade, while the percentage of schools with toilets and electricity had doubled. Primary school enrollment had increased to 96.7 percent, perhaps linked to the fact that four out of every five schools had a functioning midday meal program—four times as many as a decade ago.

But Muralidharan argued that while infrastructure had improved, learning outcomes remained bleak. He pointed to research showing that fewer than 50 percent of children enrolled in the fifth grade at government schools were able to read a simple paragraph—something they should have been able to do in the second grade. Fewer than 27 percent of third graders were able to solve a simple two-digit subtraction problem. The conclusion, he wrote, was that "the Indian education system is doing well at enrolling children in school but failing when it comes to teaching them even basic skills" like reading and arithmetic.

In the heart of Koramangala, in the southern tech-hub city of Bengaluru—formerly Bangalore—a group of engineers is plotting to fix what they consider to be modern India's greatest problem: a broken primary school system.

The offices of EkStep—*ek* means "one" in Hindi—are housed in a four-story building, tucked away in a leafy residential area. I met the start-up's

CEO, Shankar Maruwada, in a conference room on the building's top-floor terrace.

Maruwada is an alumnus of India's Aadhaar team, the group that created and implemented the world's biggest biometric identification system, with more than a billion Indians registered. Aadhaar was started by Nandan Nilekani, the billionaire cofounder of Infosys; Maruwada was one of his key lieutenants. For Nilekani and his wife, Rohini, attempting to use technology to fix education in India seemed like a suitable follow-up act to Aadhaar. They created EkStep.

"A combination of an impossible goal and an impossible time frame makes you think out of the box," said Maruwada. "The Aadhaar experience taught us that when you do that, the universe conspires to help you reach your goal."

The Nilekanis provided seed capital of around $10 million to create a technological platform that could fill some of the gaps in India's broken education system. They handed the day-to-day reins to Maruwada.

"The big problem with Indian education isn't enrollment, it's learning outcomes," said Maruwada, citing Muralidharan's research. "How can a child spend five years in school and still not be able to read or do basic math? We wanted to address that by using technology."

Maruwada pointed out that there were three scarce resources when it came to education in India: quality teachers, quality materials, and time. The third surprised me. Maruwada explained that many Indian children were pressed into work from a young age. "Do you milk the cow or study math?"

"We had to come up with something that urgently adds unique value and is not recreating the wheel," said Maruwada. He and his colleagues applied the same principles that served them well in creating Aadhaar: take a focused task and drill down. "Inch wide, mile deep," said Maruwada. "We chose to ignore fancy outcomes and target basic numeracy and literacy. And a mile deep, here, means 200 million children. That's our target in five years."

Maruwada had assembled an all-star team of coders and engineers—many of them former colleagues from his Aadhaar days—and they quickly set to work. "Can we make sure every child has basic numeracy and literacy? Can we create something to fix this?"

That something was essentially a free, open-source, smartphone app platform to create educational tools for math and literacy. The EkStep platform itself was just a starting point, with the hope that teachers across India could create their own material in their own languages, easily indexed and searchable across the country. "We think of ourselves as the creators of GPS," said Maruwada. "Let the others build Uber." The point of this was to reach the greatest number of people—especially poorer families who couldn't afford subscription-based software like BYJU's, a popular Indian mobile-learning platform. EkStep had also been commissioned by the government to create a national teacher platform to give educators the means to connect, update their knowledge, and get access to the latest teaching materials and techniques.

Maruwada called in his team to show me a demonstration of how EkStep worked. They were an idealistic, passionate lot, peppering their sentences with words like "solve," "fix," "implement," and "scale." Their demonstration of EkStep was a practiced one, but there was a palpable sense of excitement as I asked them questions. An engineer hooked his laptop up to a big screen on the wall, where he then walked us through some of the platform's uses. One could create storybooks in several Indian languages, build basic math games, develop science exercises—all using the framework of the EkStep platform. It was essentially software that made it easy to build other software. The learning games could be played on the simplest smartphones, turning small devices into miniclassrooms. Unlike PC educational tools from a decade ago, the smartphone was set to reach an unprecedented number of people in India; the race was to match the physical distribution of smartphones with greater access to free, high-quality educational content in a range of Indian languages.

EkStep's story reminded me of another experiment in India more than a decade ago, called Hole in the Wall. The project was devised by Sugata Mitra, who worked as the chief scientist at the National Institute of Information Technology, better known as NIIT. Mitra had been researching the possibilities of learning through computers. In 1999, he tested his ideas in the perfect live laboratory: a slum in New Delhi.

Mitra embedded a computer into a wall in the slum, a few feet off the ground and hooked it up to high-speed internet. He left the computer unattended.

By Mitra's account, the rest was magic.[6] Children without any knowledge of English or computing began clicking around and experimenting.

The children would make mistakes, learn a few things, and share their learning around. A sort of community knowledge began to build and be shared. As Mitra set up more of these holes-in-the-wall across India, he began seeing some surprising results. In a Rajasthan village, a group of children used a computer for the first time and within merely four hours were recording their own music and playing it back to each other. In a South Indian village, young boys had assembled a video camera just fourteen days after they first used the hole-in-the-wall PC.

Mitra's experiments turned more ambitious. In Hyderabad, he met a group of children who spoke English with a strong Telugu accent. Mitra gave them a speech-to-text software and told them to speak to it. The software couldn't understand the kids. Mitra simply walked out, saying, "Make yourself understood," and gave the children two months to do so. When he returned, the children's accents had morphed into neutral British-Indian accents, similar to the style in which Mitra had programmed the speech-to-text software.

The experiments highlighted how computers could help students learn things. In parts of the world where access to high-quality teaching was expensive or limited, computer learning could be a vital addition. Even so, the Hole in the Wall experiment faced hurdles: it wasn't profitable, and it was impossible to scale. (Professor Mark Warschauer of the University of California, Irvine, for example, has pointed out that unsupervised learning has severe limitations and can't be expected to function outside formal institutions.)[7]

I put the Hole in the Wall experiment to Maruwada. Was that how intuitive learning could work? "No way," he responded. "It's a clever solution searching for a problem. It's a knight in shining armor saying, 'Here is a way kids will solve life with tech.' But really, is that how we bring up children? Give them a toy to figure things out?"

One answer could be to give the "toys" to teachers, especially when the implements in question are cheap smartphones. Seema Bansal has been working with the Boston Consulting Group's (BCG) development team in India for nearly two decades. One of her recent tasks was to try to implement smarter teaching across a network of fifteen thousand public schools in the Indian state of Haryana. Once, a member of Bansal's team saw a local schoolteacher use sticks and stones as props to explain a mathematical concept. This excited the team: a routine math class suddenly became a fun game. Bansal wanted to get teachers at other Haryana schools to use the same method.[8] But spreading the message across fifteen thousand schools and a hundred thousand teachers proved difficult. It was a system Bansal called "The Chain of Hope." First a letter would be written at the state headquarters, addressed to each of the district offices. The hope was that in each of these district offices, an officer would receive the letter, open it, read it, and then forward it to the next level—the block offices. If it made it through those steps to the block offices, then officers there would hopefully receive the letter, open it, read it, and take the trouble to pass it on to the fifteen thousand principals. There too, if it made it to the hands of a principal, one could only hope it would be received, opened, read, and understood. In other words, it would be a minor miracle if the instructions in the original letter were implemented on a mass scale. "The solution was WhatsApp," says Bansal. None of the schools had computers or email, but what the teachers did have was smartphones. Today, principals and teachers in Haryana are divided into hundreds of WhatsApp groups. Anytime something needs to be communicated, messages are posted and forwarded. "It spreads like wildfire. There's no other way of getting things done," says Bansal. "And that's why when you go to a school in Haryana today, things look a bit different."

One of Bansal's other projects with BCG is her team's work with *anganwadis*, or rural mother and child care centers—essentially pre-Ks serving as playschool and immunization center rolled into one. *Anganwadi*s are usually staffed with female workers who have not studied past the eighth grade. "The whole point of early childhood education gets missed," says Bansal. Partnering with the World Bank and the Gates Foundation, Bansal's team is now developing an app to facilitate *anganwadi* workers as early childhood educators, empowering them with vernacular, easy-to-use smartphone games and activities for children. "This has huge potential," she says. "It could really change so much for the better."

Bansal has great hopes for EkStep too. "I have huge respect for what they're doing," she says. One of their advantages, she says, is money. Innovation in education has often been limited because it is difficult to monetize. The Nilekanis have taken money off the table by starting EkStep off with a $10 million grant and continuing to give it millions every year. "Education is risk averse," says Maruwada. "But not us. With the scope and scale of what we are doing, we can enable the government to reach every teacher in India. And through that, we can reach students in a way that was simply not possible before."

Since computer-aided learning is new, the data on its efficacy has been limited. But in a July 2017 study, Karthik Muralidharan, along with the economists Abhijeet Singh and Alejandro Ganimian, was able to find conclusive data showing that technology could help students in the developing world.[9] The paper described a lottery system to put students in a technology-aided after-school program and found that lottery winners scored significantly higher in both math and Hindi—relative to lottery losers—after just 4.5 months of access to the program. Academically weaker students made the biggest gains, again highlighting that much of their weakness stemmed from not being able to read or do simple math. The gains at a younger age could turn out to be exponential the older these students get, motivating them to keep learning and fight for higher education and jobs.

When a system is broken, technology can enable new and innovative solutions. But technology has always been expensive and inaccessible to India's poor. "That's what is beginning to change because of smartphones," said Maruwada. "And that's why our race is to provide great content worth millions of dollars and put it on the market for free."

As I listened to the EkStep engineers and their grand aspirations to help Indian children learn math and become literate, I couldn't help but ask myself if technology could succeed where all else seems to have failed. Is it possible? I imagine Abdul, for one, would say yes, it is.

3

Missed Call

The Smartphone and Job Creation

IN THE SUMMER OF 2015, the government of Uttar Pradesh began putting out advertisements looking for "peons"—the local term for low-ranking office helpers. UP, as the state is known, is home to more than 200 million Indians, packed into an area about the size of Texas (which has one-seventh as many inhabitants). Fittingly, UP needed a small army of new peons: in all, 368 jobs were posted.

A very strange thing happened next.[1] Applications poured in. After a painstaking survey that took weeks, 2.3 *million* résumés were counted. There were 6,250 candidates for each available position. Some of the applicants had doctorates.

While peon jobs are stable—even respectable—they are by no means glamorous. Peons are usually the first people one sees at Indian government offices, dressed in shabby, faded khaki uniforms; their work involves tracking down dusty files, fetching tea, and ushering in guests. Salaries range from just $150 to $250 a month.

The question is why these low-skill, low-paying jobs were in such high demand. There are several possible explanations. First, $250 a month may sound like a pittance, but it is not insignificant: it amounts to nearly double the median national salary. Second, peons are influential gatekeepers in Indian bureaucracy. If you need to see a local officer, a small bribe can go a long way. But workplace corruption is hardly something young, idealistic Indians aspire to (let alone the ones with doctorates). Something deeper was going on.

A third possibility is that India simply isn't creating enough jobs. A 2016 report by the United Nations Development Programme (UNDP) revealed that India's working-age population expanded by 300 million between 1991 and 2013.[2] But during those same twenty-two years, the UNDP says, the economy created just 140 million new jobs. Put another way, 160 million working-age Indians were without formal employment.

Job creation is the number one headache for India's policymakers. By some estimates, India needs to create a million new jobs every month simply to keep pace with the gush of new entrants to the workforce. There is little evidence that India has a plan to meet this demand. (This is the other side of the "demographic dividend" debate, where it is argued that India is facing a looming social disaster with too many young people and a deficit of jobs.)

Recruiters say part of the problem lies in the quality of job seekers. India's National Skill Development Mission reports that only 4.69 percent of India's workforce has undergone formal skill training of any kind.[3] This compares poorly with most top economies. The corresponding number for the United States is 52 percent, with an escalating 68 percent, 75 percent, 80 percent, and 96 percent for the United Kingdom, Germany, Japan, and South Korea respectively.

India's low skill levels and a shortage of full-time work have led to a chaotic national bazaar of temp workers: roadside vendors, odd-job men, domestic helpers, farmhands, and more. These workers comprise 90 percent of the national adult population, partaking in what is euphemistically called the informal sector—an off-the-grid system without work contracts, health care, or pensions. They often have seasonal jobs with no employment security. Their livelihoods are off the books. Most don't have bank accounts, and they don't pay income taxes. (In any case, only about 5 percent of Indians *qualify* to pay income tax in India. On paper, 95 percent make less money than the required government cutoff.)

It is a rather bleak outlook. A shortage of jobs threatens to upend India's growth story.

This is where the mass adoption of smartphones could be a positive development. Other countries have already created new types of jobs through what is known as the sharing or gig economy. In India, this is belatedly taking off. Ride-hailing apps like Uber and Ola create hundreds of thousands of flexible—and air-conditioned—jobs for drivers; UrbanClap creates a location-based marketplace for quick one-off tasks, similar to TaskRabbit in the United States; Flipkart, Amazon, and Snapdeal are creating a national infrastructure for packing and warehousing, and building trust in a burgeoning system of deliveries; Airbnb and OYO are organizing the marketplace for accommodations; the likes of Swiggy, Zomato, Foodpanda, and Caviar are expanding business for restaurants—and deliverymen—by creating a food-delivery culture; and so on. India is starting small but has all the market conditions to thrive in the age of the gig economy: large numbers of temporary workers, insufficient availability of traditional services, and a population young enough to embrace technology.

For better or worse, smartphones could well hold the future of Indian work.

A few years ago, when I was living in New York City, a friend excitedly told me about a new smartphone app that would change the way I traveled. It was called Uber. The friend recounted to me how he had used it on a date. He ordered a black Lincoln sedan right to his doorstep. When it arrived, he escorted his girlfriend to the car. The driver opened the door for them. The car was clean and comfortable, with plush leather seats. When they arrived at their restaurant, there was no need to fumble for cash or swipe a card. They simply thanked the driver and walked away. It was subtle and classy, like being chauffeur driven. It was a taxi without the hassle. (Incidentally, the date was a success. The couple got married the next year.)

Uber seemed to fit what I needed. I disliked carrying cash; I wasn't very good with directions; taxis were hard to come by on my street. I ended up using the service frequently in New York City. The cars would arrive quickly, with just a few taps of a fingertip. The directions were always precise and clear on GPS. New York was a well-mapped city.

I had a different experience in New Delhi. Opening the app is easy: that's the same everywhere. After that, India begins. When I tried ordering an Uber to my apartment one time, it turned out my home address wasn't listed on the map. Not to be deterred, I homed in on a spot that was precisely where my apartment was—I could even see an outline of the building—and set it as the pickup location. A driver was assigned to me. He was fifteen minutes away. It was not the usual two or three minutes I had gotten used to in New York, but still, I was ready to wait.

Given the sudden surfeit of time, I decided I would have a shave. I went to the washroom and lathered up. Right then, the phone rang. It was the driver.

He spoke in colloquial Hindi. "Hello-ji. Where are you?" he asked.

Me, also in Hindi: "Um . . . home?"

Him: "How do I get there?"

Me: "Can you follow the map?"

Him: "*Map-wap* doesn't work. I'm in Noida. Can you give me directions?"

I wished I could. I didn't know the way from Noida, which was technically in a different state. I was new in New Delhi.

Me, mildly embarrassed: "Sorry, I don't know the way. Can you just ask around, please? Ask for Humayun's Tomb in Nizamuddin East?"

When you live in India, you learn to have a "landmark" to offer people as directions to your home. In my case, it was a giant Mughal-era tomb, a predecessor of the Taj Mahal. My driver had never heard of it.

Him: "OK, OK, let me try." He hung up.

I sighed and lathered up again, this time with warm water and more flourish. I picked up my razor.

The phone rang again.

"Hello-ji," said the driver. "Lots of traffic."

"OK," I said. "Good to know."

"Thank you, ji," he responded. "I'm coming."

"OK, thank you." I hung up, bemused by the updates.

A little later, as I was washing up, the phone rang again.

"I'm still coming," declared the driver.

Silence.

"Yes, why not?" I asked. I was beginning to get frustrated now. One of the things I enjoyed about Uber in New York was being able to book a car without having to talk on the phone.

"Well, it's been fifteen minutes," he said. "I wanted to make sure you were still waiting for me." It was clear that as much as I didn't trust him to make it to my place, he didn't trust me to stick around. This was almost like a game of chicken.

"Oh. Yes, I am. Otherwise I would have canceled. When will you get here?"

"I'm almost at Nizamuddin. How do I find your place?"

I gave him directions. "Yes, yes, left at the petrol pump."

"PETROL PUMP? HAANH?" The driver may have inadvertently gone on speaker phone at this point.

"Yes. PETROL PUMP!" I began yelling too. "Then left, then straight, down along the wall, all the way. Yes, LEFT."

"OK, OK, don't shout. I told you I'm coming!"

He missed the left turn. I could see his car move on the map, like an ant feeling its way around Delhi, stopping every now and then as it adjusted its sense of direction. It was another ten minutes—and two more phone calls—before my hapless driver finally arrived. I had to stay on the line for the final, tortuous five minutes, from my balcony, peering outside to make sure I spotted a white Maruti car. As it turned out, my address, B-19, was somewhat separate from some of the other Bs. Much of the neighborhood was planned in a haphazard fashion, with the As leading to Cs, and the Bs to Ds.

When I got into the car, finally, I felt ready to burst. It had been nearly half an hour. I was late for work. But when I spoke to the driver, my anger melted. He had just moved to Delhi a few months ago and was unfamiliar with the roads. Didn't he know how to use his smartphone? No, he didn't. He couldn't understand maps. He couldn't read English. He couldn't even figure out how to press the button on the Uber app to start our trip.

I helped him out. We pressed Start together. I navigated for him. My Uber India journey was finally under way.

"Delhi has twice as many cars as it did ten years ago," pronounced an American man in a crisp white shirt and black trousers. "But guess what? Delhi might be just five or ten years away from no traffic at all."

The speaker was Uber's founder and CEO, Travis Kalanick. He was addressing a small gathering of Delhi's business elite at the city's plush Taj Mahal Hotel. It was a cool December evening in 2016. Delhi attracts tens of thousands of tourists in the winters. The city's arteries were seized up with traffic—more than the usual, which was bad enough. To add to that misery, pollution was causing serious disruptions. Schools had recently shut down because the air quality was deemed hazardous. The skies were heavy with smog. Visibility was low. Being out and about—especially stuck in traffic—had begun to feel hellish. Kalanick had picked a topic he knew would resonate. He was explaining the benefits of Uber's new carpool service.

"It's not really about the car itself. It's about how we use it. Just think about private cars—we use them for just 4 percent of the day. So these large hunks of metal are in storage for 96 percent of the day. What other kind of asset gets so underutilized? Twenty percent of the land in a city is used to store these vehicles. That's land that can be used for more housing!" declared Kalanick, in what I took to be a well-rehearsed set of revelations. "Now imagine what we can do if everyone's taking an UberPOOL ride. We're using the cars 100 percent of the time, and fewer cars on the roads. It creates a mass transit system that goes to everybody's doorstep."

Kalanick wasn't just being altruistic. A few months earlier, Uber had been forced to exit China, a market where it was burning through an estimated $1 billion a year. Uber sold its national operations there to its bigger local rival Didi Chuxing, a company whose name literally means "honk, honk, commute."

Kalanick saw India as ripe for an Uber expansion. This was his big bet. The timing seemed perfect, with millions of young Indians simultaneously getting online and moving to the cities. Only 3 percent of Indians owned cars—a tiny ratio when compared with China's 14 percent or America's 79 percent. This made the potential market for Uber relatively vast. Public transport in India was poor; middle-class Indians were desperate for alternatives; India had a more open market than China's. (Not too long after his India visit, Uber's board forced Kalanick to step down as the company's CEO after a spate of reports about workplace harassment. The move had little impact on Uber's India plans.)

There were, however, significant hurdles to Uber's operations in India. Traffic was crippling, making rides less profitable; cities were not adequately mapped; drivers often didn't know how to use smartphones. Uber also had a giant Indian rival in Ola, which claimed to operate in four times as many cities, with more than double the number of drivers. As a local player, Ola had been canny in understanding the market's needs. It had introduced cash payments as an option to settle fares, and the move had become immensely popular—more than 80 percent of its ride transactions were in cash. It was also the first to create versions of its app in Indian languages to attract more drivers.

Even though I had my own car in India, I took countless Uber and Ola rides around the country. In fairness, the Uber India ride I described earlier was probably one of my worst. My experiences varied. Sometimes, the drivers were adept with smartphones and maps. Mostly, they were warm and friendly. But there were often other problems. On several occasions, cellular internet dropped out, so navigation was impossible. The pickup experience continued to begin with at least one phone call; drivers would ring to check on my location or ask for directions. We struggled with simple words that wouldn't translate. A hotel porch or lobby, for example, remained difficult to explain in Hindi, especially when the driver had only recently moved to the city and couldn't imagine what a large hotel building looked like.

Chatting with Uber and Ola drivers afforded me hundreds of data points about the health of the ride-hailing industry in India. There was an instant ease to these conversations. I found that Indian drivers would willingly discuss their lives, often in intimate detail, and with no shyness about salaries and numbers.

The numbers, however, made me sit up and think. They did not seem credible. Several Uber drivers told me that they were making around 80,000 rupees a month (about $1,200). This was a wildly lavish sum for Indian drivers in 2014, putting them just shy of the top 1 percent of earners nationally (according to the economists Thomas Piketty and Lucas Chancel, the threshold for the 1 percent in India is an annual income of 1,226,689 rupees, or nearly $19,000).[4] By comparison, most private chauffeurs would make about 16,000 rupees a month. Regular taxi drivers could expect to make maybe 20,000 to 25,000 a month, but working nights, weekends, and unpredictable hours. Suddenly, everyone wanted to drive for Uber or Ola: imagine making four times as much as your peers.

But the driver incomes didn't add up. After all, the customers weren't footing the bill. If I took a regular taxi from my home to the airport, I would be charged about 800 rupees. An UberX would cost me a third of that. Both Uber and Ola were subsidizing rides to grow their respective market shares. Yet the drivers kept making big money through incentive schemes. Dozens of Uber drivers told me that during 2014 and 2015 they stood to make a flat 3,000 rupees a day if they hit a certain number of rides during peak hours. In other words, if they worked long enough, and efficiently, they would hit the daily jackpot. Three thousand rupees, or forty-five dollars, is an almost princely daily wage in an Indian city—it is greater than the median starting salary for graduates of the Indian Institutes of Technology. The news of this daily incentive program seemed to spread like wildfire across the country's tea stalls, *dhabas*, even universities. Thousands of recruits were signing up to become app-based drivers every week. Many took out loans to buy cars. Some drivers revealed they were recent engineering university graduates. Apparently, information technology jobs were drying up (or their skills weren't up to scratch); the real money seemed to be in driving for Ola or Uber.

Meanwhile, the size of these ride-hailing apps grew and grew. By early 2017, Uber India president Amit Jain told me, around 250,000 drivers were regulars on the Uber platform. An additional 150,000 had driven for Uber at least once and would return to it sporadically: these were temporary, seasonal workers.

Ola's network was even larger. Founding partner Pranay Jivrajka told me it had 650,000 regular drivers, with many more who would come online on occasion.

As the companies grew, their belts tightened. The driver incentives were first reduced and then stopped altogether. Since the cost of the rides remained low, this had a dramatic impact on the take-home pay of the drivers. In 2016 and 2017, a growing chorus of drivers began to complain they were making less money. Their monthly incomes were now in the range of 30,000 rupees—still significantly higher than the incomes of other drivers, but nearly a third what it once used to be.

Sudhir Prasad was one of several drivers who willingly volunteered the numbers for me. "Please write about this," he implored, when I told him I was a journalist. "We're helpless!"

In January of 2017, when I rode with Prasad, he described how he was struggling to make ends meet. Two years earlier he had been making 70,000 rupees a month—more than twice the income he was now bringing home. At the time, he delayed paying off the loan on his car, instead renting a larger house. He moved his son to a better school so he could study English. Now he was in debt.

"Why are they playing with our lives? They got us used to all these incentives, and then suddenly they took it all away."

I tried to explain how the companies likely had an algorithm that set the daily bonus targets.

"Do they think our lives are a computer game?"

Prasad was hardly alone. Anger was building among thousands of app-based drivers who said the promises of big money had suddenly dried up.

In February of 2017, things came to a head. Tens of thousands of Uber and Ola drivers went on strike across India. They wanted their daily target incentives back. The drivers asked for a guaranteed minimum wage and benefits. Cities like New Delhi, Mumbai, and Bengaluru were badly hit. Tens of thousands of city workers had come to depend on the cheap ride-hailing apps to commute to work; now they were stranded.

As the strike extended into a second week, I couldn't help but mark the irony. While the drivers felt hard done by because they were making less money, Uber and Ola were themselves running losses in a race to grow their market shares. The ride-hailing industry seemed like a giant Ponzi scheme.

Labor disputes always end with one side blinking first. The drivers gave in. They needed the work. The drivers' loan sharks, it turned out, were less forgiving than the ones backing Uber and Ola.

When I met Uber's Amit Jain for an interview a few weeks later, he told me the strikes had not even been organized by drivers who were on his platform. "Think about it this way," he said. "If you're an Uber driver, how do you mobilize and stop tens of thousands of other drivers from working? You can't. You don't even know them. The protests were organized by professional union people who wanted to bring down the ride-hailing-app industry."

Even if that were true, wasn't Uber raising driver expectations, giving them a taste of the good life for a bit, and then bringing them crashing back to reality?

Jain was unflappable. "The opportunities we provide have to be more attractive than other jobs on the market. If the earnings aren't good, if they aren't stable, drivers won't join us," he said. "But they are."

Indeed, drivers—each needier than the next—kept lining up to work for Uber and Ola. Jain said his aspiration was to grow Uber India's pool of regular drivers from its current strength of 250,000 to a million by

2019. "Just look at it this way," he told me. "The Indian Railways has been in existence since the 1850s. After being in business for 167 years, they employ 1.3 million Indians. We could be providing jobs and livelihoods to a similar number of people in just five years of operating in India."

He paused. "Still think we're not creating sustainable opportunities? There is a great demand for this work."

Ola's Pranay Jivrajka was even more bullish about his company's ability to create work. "We intend to create 5 million opportunities in five years," he said, careful not to use the word "job," which of course connotes security, pensions, sick days, and benefits. "It's not just about creating jobs," he explained. "It's about creating an ecosystem.

"For most other skills, first there is a demand and then people enroll in a training program, get skilled and certified, and then find a job. With ride-hailing apps, there is an inherent demand, which is constantly increasing with supply. Supply drives more demand—this is unique in economics," argued Jivrajka.

When I pushed him on the collapsing driver incentives, and whether he saw that as unfair to a generation of drivers who had gotten used to higher pay, Jivrajka opted to shift the blame elsewhere. "Drivers suddenly started earning way more than we expected because of a land-grab and monopolizing attempt," he said, without naming the Uber-sized elephant in the room. "Our opposition was adopting a global cookie-cutter approach. That's what has gotten scaled back, and that's why we're back to reality now."

Sudhir Prasad, who rode a wave of heightened expectations in 2014, and then a collapse in 2017, was scaling back too. He told me he was thinking of moving his son back to a cheaper government school and finding a smaller home to live in. I asked him if there were any other jobs he could do. "What could I do? I used to work in the army, and I retired early. The good thing about driving a car like this is that it's respectable, unlike the regular taxi services. I sit in an air-conditioned car all day. In this heat and pollution, it's a godsend. I've gotten used to it. Now, this is all I can do. I'm stuck. The money is still better than other jobs I can find. What other skills do I have?"

Prasad's dreams were collateral damage in a multibillion-dollar turf war, waged in India and backed by investors in Tokyo and Silicon Valley. When you weigh Prasad's anxiety with a potential 5 million new job opportunities, you have to ask whether it's worth it. Is this India's messy, disruptive drive forward?

•——•——•

By some accounts, India's greatest source of new jobs will be the burgeoning e-commerce industry. Brick-and-mortar bazaars are fast moving online as giant e-commerce companies like Amazon, Flipkart, and Snapdeal compete for customers by offering scarcely believable discounts. "Holiday bonanza—50% off washing machines! Men's clothing—80% off! Sofa sets, buy-one-get-one-free!" The discounts are designed to slake an innate thirst for bargains. Internet shopping offers the satisfaction of the bazaar deal without the hollow feeling of hustling a struggling shopkeeper.

As more Indians buy things online—mostly on their smartphones—an army of workers will be needed to stock, pack, and deliver internet purchases. When Ramesh in Coimbatore orders a bag of toilet paper on Snapdeal, where does it come from? When Smriti in Surat clicks Buy on a new table lamp on Amazon, how does it get to her? E-commerce platforms rarely own an inventory of products. Often, they simply create a virtual marketplace and act as the middleman. That's where companies like Vulcan come in, specializing in the logistics of mass storage and delivery. (Vulcan is owned by a company called Jasper, which in 2017 also owned the Indian e-commerce marketplace Snapdeal.)

The first time I stepped inside a Vulcan warehouse, it was like a startling vision from the future becoming reality. Row upon row of fifty-foot-high racks, as far as the eye could see, stocked and stacked full of smartphones, TVs, fridges, microwaves, washing machines. Each megarack had a conveyer belt going from top to bottom, winding its way down like a water slide at a theme park, stopping at each shelf. Once a product was selected, it was automatically placed on the conveyer belt, journeying down. It was impossible not to marvel at the system. The only thing I had ever seen like this was the soda and candy vending machines in the United States; the warehouse took a similar concept and applied it to a bewilderingly larger canvas.

The Vulcan warehouse I visited was in a village near Najafgarh, a dry, underdeveloped town south of New Delhi. A giant cuboidal building housed a hundred thousand square feet of workspace. The warehouse functioned like a U-shaped conveyer belt. At one end, products were received, sorted, marked, and then moved inside for storage. At the other end of the U, packages were retrieved from storage, packed, and then marked for delivery. Thousands of workers made this happen. One group placed products on a series of long conveyer belts; another stocked and retrieved; yet another group processed invoices; the largest set of workers was the packers, taping and cling-wrapping cardboard boxes of all sizes in a constant hum of screechy sounds. It was a well-run supply chain. This was where India's aspirations were quite literally stockpiled.

Babloo Kumar was convinced e-commerce would keep growing. He was twenty-two, a recent college graduate, tasked with making sure that each package sent out for delivery had the right products inside it.

"This is India," he declared, beaming. "Everyone wants the latest smartphones," he added, pointing to the packages that were being shipped out. Most of them tended to be cheap, $50 to $150, he said. He seemed happy to practice his English with me.

Kumar was making just 20,000 rupees a month, or about $300. But this was a large, generous sum of money for him. (The salary, while seemingly low, is high enough to place Kumar among the top 10 percent of wage earners in India.) His father was a priest at the local temple, a high-caste Brahmin who would recite religious hymns for hours every morning, for little pay. "I want to work in the real India. No jobs in my village," he sighed, smiling ruefully. "Growth of e-commerce is creating all these jobs for people like me."

Kumar's assertion is one side of a vigorous global debate as to whether technology will create more jobs or destroy them.

The question is an old one, as Erik Brynjolfsson and Andrew McAfee argue in *The Second Machine Age: Work, Progress, and Prosperity in a Time of Brilliant Technologies*. As early as 1811, English textile workers panicked at the prospect of automated looms and attacked the mills they worked for. In the 1930s, John Maynard Keynes popularized the term "technological unemployment" to describe how machines could render labor useless. But, as Brynjolfsson and McAfee point out, a general view was formed that automation created more opportunities than it destroyed. That, however, happens during what the authors call the First Machine Age. The Second Machine Age could be different. That's when human brilliance leads to the invention of artificial intelligence and driverless cars, of robots that can perform everything from warehouse work to surgeries. This would not be solely a first-world problem. "In the long run, the biggest effect of automation is likely to be on workers not in America and other developed nations, but rather in developing nations that currently rely on low-cost labor for their competitive advantage," they wrote. For a high-population nation like India, the elimination of low-cost labor could be especially debilitating—both for jobs and competitiveness.

Another problem is the sheer pace of innovation. In the past, technological revolutions took place over the course of decades, with enough time for people to retrain or to create different types of jobs. The growth of modern AI and robotics could be coming too quickly for existing businesses to deal with. Yet perhaps the conditions in India are a bit different. If driverless cars are maybe a decade away from popular use in the West, they are probably twenty-five years away from attempting to wind their way around India's potholed, chaotic roads.

HSBC's chief India economist, Pranjul Bhandari, argues that e-commerce could become the answer to India's jobs problem. In a 2016 research paper titled "More Jobs per Click," Bhandari points out that e-commerce jobs closely match India's skills and entrepreneurial profile.[5] By her forecast, India needs to create 80 million jobs in the next decade, while the economy is only set to create 60 million. In other words, there will be a shortfall of 20 million jobs. E-commerce could fill more than half of that gap, posits Bhandari, with 12 million net new jobs created in logistics, delivery, customer care, and IT services.

"If you move from just brick-and-mortar jobs to digital, then you could lose jobs," Bhandari told me. "E-commerce has better logistics and is better organized. But I argue that it's wrong to assume the consumption pie will remain stagnant in India. If I want to read a book, I go to a store nearby. If I find it, I buy. If I don't, I don't. But with e-commerce, you can buy it later. The consumption pie can grow so much, and so it will employ a lot more people. It will add to jobs rather than subtract." Not only that, but e-commerce will also allow millions of families in rural parts of India to access goods and services that are only available in the cities; the market already exists, but it is untapped.

Snapdeal CEO Kunal Bahl, perhaps understandably, agreed with HSBC's prediction on job creation. "We have an unlimited population of workers who have barely passed eighth grade," Bahl told me. "They can't get office jobs. It's easy to imagine they would have remained unemployed or become beggars. But if you give them a smartphone and software that clearly explains what they have to do—pictorial descriptions with arrows—then they can work in supply-chain businesses or as delivery men. The smartphone gives companies confidence to enforce checks and quality control. It has unlocked the potential of a lot of Indian jobseekers."

Bhandari's HSBC report released right when India's media was asking whether there was a bubble in the e-commerce market. The biggest players—Snapdeal, Flipkart, Amazon—were leaking money. Smaller internet-based start-ups like Grofers and MyButler, which depended on food deliveries, were considering shutting shop.

"But that's why we need to put it in context," said Bhandari. "The companies that succeed become exceptionally large, and they change our behavior. Ninety percent of India's retail is unorganized. E-commerce will bring a lot of convenience. Instead of running around as much you can be efficient. People will consume more."

HSBC is forecasting that with greater smartphone penetration, a youthful population, and new online payment systems, Indians will increase their online spending by orders of magnitude: from $21 billion in 2016 (2 percent of total consumption) to $120 billion in 2020 (8.8 percent of total consumption) to $420 billion in 2025 (16 percent of consumption). It's a bullish prediction, which would also require an expansion of the country's middle class. If realized, a twentyfold increase in the size of the e-commerce market would then translate into millions of new jobs in warehousing, packing, and deliveries, while also expanding opportunities for businesses.

Much of the growth, according to Bhandari, would come from a still-untapped rural India, a process akin to what happened in China over the last decade.

In 2013 the Chinese e-commerce giant Alibaba set up "Taobao villages"—rural digital hubs that put small local shopkeepers, selling everything from clothes and furniture to pottery and silverware, on the e-commerce platform. The system had two impacts: local merchants became better organized, and they expanded their potential markets nationwide. Spotting the opportunity, third-party logistics providers stepped in to create distribution networks.

By 2015, just two years in, the number of Taobao villages grew from 20 to 780. Bhandari said the movement had created millions of new jobs, simply by boosting the size of the potential market and making potential sellers more visible, targeted, and efficient.

When it comes to the internet, India usually lags China by seven years, asserted Bhandari. "Even with half the growth China is witnessing, India could have 5 million new village merchants in the next decade."

No one pays for a regular search on Google or Bing. Or for looking up directions on Google Maps. Or for browsing Wikipedia.

Each of the above is providing you a vital service while saving you time. Instead of visiting the library, you have a search engine; instead of stopping and asking people for directions, you can simply look them up online. There is a value to these services, but there's no real way of quantifying that price: it doesn't show up on GDP calculations.[6] There are few good measures to understand how technology improves our overall productivity and efficiency.

The MIT economist Eric Brynjolfsson—coauthor of the aforementioned *Second Machine Age*—attempted to measure the value of free services such as Google, Wikipedia, Facebook, and YouTube. In 2012, Brynjolfsson estimated that the US economy would have likely gained $300 billion in GDP—about 0.23 percent of additional growth over a period of ten years—because of increased efficiency and productivity from free internet products.[7]

If the findings are accurate, India would benefit in a similar way. And perhaps more so, given the dramatic inefficiencies in everyday Indian life. "We're cutting out the middleman," says Ishaan Gupta, coauthor of a series of reports known as *Digital Desh*—*desh* being Hindi for nation.[8] Gupta spent three years traveling around India putting together examples of men and women whose businesses had become more effective once they got on the internet.

"The single biggest benefit to small-scale Indian entrepreneurs is being able to connect directly with their buyers," says Gupta. "Earlier there was either an information deficit or too many middlemen taking cuts along the supply-chain route. Access to the internet is making the creator and the consumer more important, while cutting out the guy in between. This will improve products, supply chains, and business as a whole."

Gupta describes how small rural entrepreneurs in Aizawl, in the country's forgotten Northeast, had begun to find markets for their handlooms in mainland India after attracting interest on Facebook and Instagram.

"It's not easy for every small dealer to get on Amazon," he says, pointing to strict guidelines for photography of products. "But just having a smartphone allows them to connect with a wider audience. Creators, craftspeople, any kinds of producers in rural India are finding that a simple smartphone can grow their markets."

Gupta is something of an evangelist when it comes to the internet in India. He found that he could connect traders from the North and West to others in the South and East so they could compare notes on how to expand their sales.

One reason Indians have taken to the messaging app WhatsApp, for example, is the fact that one's username is the same as one's mobile number—something that Indians are already adept at sharing around with people. Indians tend to share images, data, and messages on WhatsApp instead of trying to use email, for which one needs to register and create a new username and password. WhatsApp in India—which has more than 200 million users—is also the way shopkeepers build distribution lists to let their customers know about their latest wares. It's easy to use, it's free, and it can include images and videos; most of all, it is a quick download on a smartphone app, with no need for a PC or laptop. (In 2018, WhatsApp added a feature to facilitate payments that is fast becoming popular.)

One doesn't need to travel very far in India to find business beneficiaries of WhatsApp. Near my own home, where I shop for vegetables at Delhi's Bhogal market, I found Deepanshu Maurya, a small roadside vendor. Maurya's vegetable store is a five-square-foot operation in a busy marketplace. Mounds of vegetables surround Maurya as he sits perched on his store table. He had been using a smartphone for only a few months when he chanced upon WhatsApp. He began adding his customers to a group and sharing pictures of his produce every morning. He found that instead of coming to see him, his customers could just send him messages detailing what vegetables they wanted. Maurya would then have them packed and delivered—he could afford to hire assistants now. His business has doubled. He has also figured out how to list his tiny shop on Google Maps.

Maurya's doing the same job, but the smartphone has boosted his prospects. The question is how to measure the progress of other Mauryas, scattered across the country. And, specifically in an Indian context, how do you measure the difference technologies are bringing to middle-class lives?

One place to start with is the so-called missed call, which Indians have been using for years.

When cell phones arrived in India in 1997 I was a teenage schoolboy in Calcutta. My father bought a Nokia 5110 handset and an Airtel cellular connection. For a while, it was the only mobile phone in our family; it was a prized novelty, a glimpse into the future. We would take turns trying it out. It had a basic keypad. It also had an alarm, a calendar, text messaging, and, of course, Snake. (I take little pride in saying I was a self-declared Snake champion, with hundreds of hours frittered away.)

Phone use in 1997 was expensive—especially for Indians. While a text message to another mobile number would cost just a single rupee, a phone call would cost as much as twenty rupees a minute. (With the exchange rate at the time, that would be around sixty-seven cents today.) It was exorbitant: about twenty times as much as a regular phone call from a landline. In those days, twenty rupees could buy you a large sack of potatoes or a liter of gas. On the phone, those twenty rupees were charged even if you didn't talk for the full minute; if the call was picked up, even for a second, you were billed. If you talked for a minute and one second, you incurred the fees for two minutes. And so on.

That's how the "missed call" was born. You dialed a number, and before it was picked up, you aborted the call. No cost incurred. Why would you do this? The most obvious use was to exchange numbers. "Tell me your number, I'll give you a missed call." It was the fastest way for both parties to register and save phone numbers.

"Missed call" was a noun *and* a verb. It was a hope: "Will she missed-call today?" Then you could call back from a cheaper landline.

The missed call was highly versatile. Calling someone could mean "I'm here, I've arrived on time." It could be "Hi, I miss you." It could be "I'm ready to come home. Can you pick me up?" It could mean anything. It was telephony, but it was also telepathy. Most important, it was free.

The missed call has become such an important part of everyday Indian life that today it seems completely unremarkable, quotidian. Only when I left India did I realize that it was almost unique to South Asians. (Some Kenyan and Filipino friends also seemed familiar with the concept. In the Philippines, for example, *miskol* became a popular word in the late 1990s.)

Many Indians, especially those who are going to read this book in English, have likely gotten postpaid cellular connections with call charges that are almost free. But the missed call remains important. Most Indians are on prepaid cellular connections, where each outgoing call incurs a cost. To avoid that cost, however small, the missed call thrives. If you get into an auto-rickshaw in India and ask for the driver's number, he's going to give you a missed call. Picking up—and incurring him a cost—would be bad form. (It's rather more likely that he will ask *you* to give *him* a missed call—that way he can personally ensure the call isn't picked up. As we've established, India is a low-trust society.)

Even in the era of smartphones and data, the missed call remains an important part of Indian life and business. Banks can receive a missed call from a registered number and text you back a mini account statement. Political parties ask citizens to give them a missed call to get text updates about rallies and protest locations. Prime Minister Narendra Modi has set up a number for people to missed-call in order to receive a free callback with an audio recording of his monthly radio talk show. In fact, if you have a basic feature phone, you can use the missed call as a radio service, by calling a music channel's number, after which an automated system calls you back to play music and advertisements. You can join mailing lists for products, TV shows, music competitions, all by giving specific numbers a missed call. You can sign up for jobs, for events. There is only one thing all these calls have in common: they're free. Indians love a good deal.

When I first met Sean Blagsvedt, at breakfast in Bengaluru in early 2017, he seemed flustered. He had just dropped off his son at school, then gotten a call from the teachers. His son had left his glasses at home.

Blagsvedt was in his late thirties, with a wide smile, an athletic build, and a ponytail that made him look like a character from *The Karate Kid*. As we sat at our table ordering coffee, he apologized for being on his smartphone.

"Have you heard of Dunzo?" Blagsvedt asked, fingers tapping away on his keyboard. "I'm getting a guy to go home, get the glasses, and drop them off at my son's school."

In Bengaluru, the land of apps and new technologies, Dunzo was the very latest wonder, spoken about in hushed tones of awe. Type in your request, and someone would be dispatched to make your life easier: deliver a pack of cigarettes, get your TV fixed, let your plumber in. And you could do this all without speaking to anyone on the phone. All you needed to do was open the app and type out your request.

Blagsvedt looked up finally, apologizing for being distracted. He had found someone to run the errand.

We were meeting to talk about Babajob, Blagsvedt's Bengaluru-based start-up. As the name suggested, the company's aim was to target India's job market and make it more efficient.

"In India people usually get jobs because of their social networks—by which I mean *offline* social networks," said Blagsvedt. "If we had LinkedIn for the village we would be set. Everyone would have a job." Blagsvedt's belief was that there were enough jobs in India, and certainly enough job seekers—but that there was no efficient way to connect the two. The internet couldn't bridge this divide just yet; not enough people were online, and new users weren't tech-savvy. There was an immediate need for options available to the 800 million Indians still without the internet.

Babajob is a blue-collar version of LinkedIn. It has a simple concept, beginning with a job seeker giving the Babajob hotline a missed call. A digital system will then call the job seeker back, in a language of his or her choice, seeking simple yes-no or multiple-choice answers to build a profile of the worker and the type of job he or she is looking for. A short while later, the Babajob system will send a text message to the job seeker with a job option, salary, location, and contact number.

For Blagsvedt, making the introduction is most of the job done. "If you show up for a delivery-person job interview at Domino's, the odds of you being hired are 90 percent," he said. "Ninety percent! For low-end jobs, if you qualify for the screening criteria and you show up, that's enough to get you started. It's enough to get you in the door and get you paid that week. The jobs are there—you just need to find them."

Babajob has about 7.5 million job seekers on its platform, and it gets fifty thousand missed calls a day. Salaries for the jobs on offer tend to be low—around $200 a month—but demand is high. The jobs tend to be mostly urban: drivers, maids, deliverymen, warehouse workers, packers. Babajob isn't creating the jobs; it's connecting job seekers with large companies already looking for workers. "The connection from employer to employee was constrained by the size of one's offline social network," said Blagsvedt. "Babajob closes that gap for blue-collar workers."

Blagsvedt has been in India since the early days of the information technology boom. In 2004, he left the United States for Bengaluru and took a job in Microsoft's regional research laboratory. Some of his early research explored what technology was doing for people who weren't computer users—the economic impact of a phone for a flower seller, for example. "They could suddenly process many more orders," said Blagsvedt. "The productivity gains were immense, just because of this one small device."

Blagsvedt made India his home as he fine-tuned his idea for a start-up. "Just as the PC made things radically efficient in the white-collar space, I figured the phone would do the same for the blue-collar space. And if you get a smartphone, then it only speeds up that process."

We were interrupted by a call on Blagsvedt's phone. It was his wife. She was asking why a deliveryman had showed up at their doorstep. "It's Dunzo, honey," he explained. "He's come to get the glasses!"

Blagsvedt's face would light up in these moments. The notion of technology fixing a problem was something he had given a lot of thought. He explained how he was influenced by the Nobel Prize–winning economist Amartya Sen's book *Development as Freedom*. "Why does exploitation take place? In part, it's because opportunities aren't abundant," he said. "If you look at caste, or how real economic opportunities here are limited to the privileged, those are important things to try and disrupt."

Blagsvedt is not the only American technologist to come to India and chance upon the potential of the missed call. One of the very first people to base a business idea around the concept was an American called Valerie Wagoner, who cofounded ZipDial in 2009. Wagoner at the time was looking at ways to help retailers track consumer loyalty and increase brand engagement, a tricky task given how Indians mostly shopped at small corner stores and paid for goods in cash; there was little available data on consumer behavior and communication with buyers. Then Wagoner discovered what missed calls were—and how well they matched the Indian consumer's psyche. It felt like a gold mine.

One of her first experiments with missed calls was during the 2010 FIFA World Cup: fans could call a number and get free texts with score updates. When the experiment was repeated for cricket, the number of users exploded. Millions of Indians were dialing—and hanging up—to get free score updates. It was all word of mouth; Wagoner hadn't spent a penny on marketing costs.[9]

ZipDial was born, with large clients like Unilever, Gillette, and GlaxoSmithKline signing up to create campaigns where customers could watch an ad, dial a missed call, and then get a return call introducing them to a lucrative client referral program. (ZipDial was acquired by Twitter in 2015 for a reported $40 million. And, as often happens in this industry, Babajob was acquired in late 2017 by its bigger Indian rival Quikr.)

In the months after I met Blagsvedt, I spoke with several Babajob users. One of them was Renu, a forty-three-year-old cook in New Delhi. Renu had three children and needed to work to support them. When I met her, she had taken a thirty-minute break from one of her jobs, at a diplomat's home. She was cooking dinner for a party of eight. She explained to me how hard it was to find a job these days. When she moved recently from a different part of the city, she had lost many of her local connections. She didn't know where to begin asking for work. Babajob had helped her connect with rich Delhiites who needed cooks at home. But she wasn't happy with her current job, she said. "They don't like my food so much," she confessed, in Hindi. "I'm going to call for more jobs soon. Now it's just a missed call away. And the best part is that it's free."

PART TWO
Society

4

Lord Kāmadeva's Digital Bow

Dating and Marriage

WHEN SIMRAN ARORA RETURNED to New Delhi from London, master's degree in hand, her parents welcomed her with an enough-is-enough ultimatum: she was twenty-six, and it was time to settle down with a good Punjabi boy of their choosing.

"I said sure, why not," recounted Simran, four years older (and wiser, as I was to find out). "If the guy is Mr. Right, who cares if it's an arranged marriage?"

Simran isn't her real name. She asked me to keep her identity secret because she didn't want her family and friends to learn the details she was about to tell me. "It's a complicated, messy, crazy story," she warned me.

Simran's willingness to be matched by her parents was not unusual. The 2012 India Human Development Survey found that a mere 5 percent of women picked their own husbands; 22 percent made their choices along with their parents or other relatives, and 73 percent had their spouses picked for them with no active say.[1]

When marriages are "arranged," parents usually filter candidates based on compatibilities of caste, class, and family. In many cases, the stars must be aligned—quite literally—as astrological charts are matched to ensure a future of marital harmony.

Not everyone follows convention. A small but growing number of Indians, mostly young urban professionals, dismiss the prospect of being set up. Their alternative is the curiously named "love marriage"—a union that implies not only the serendipity of falling for someone but also a proactive, defiant choice. Adding the prefix "love" attaches a hint of illicit romance to what is known in most other parts of the world as, simply, marriage.

The choice isn't always binary. Sometimes unions nestle between "arranged" and "love." There is, for example, the increasingly common "arranged-to-love" approach, where old-school-but-liberal parents allow a family-matched couple to go on several dates in the hope of Cupid doing his thing. (Incidentally, Indians have their own version of the Greek god: the Lord Kāmadeva is often depicted as a handsome man with green skin, wielding a sugarcane bow with a bowstring of honeybees. Simultaneously stinging and sugary, love is indeed a bittersweet affair.)

Simran was a pragmatist. She didn't care for the reckless pursuit of love. What was the point? Instead, she trusted her parents to find her a man who would be a loving partner for life. "Marriage is about family, not just love," she says that she thought at the time. "My parents' opinion counts."

The Aroras quickly got to work, doing what many of Delhi's upper-middle-class families do in these circumstances: they hired a marriage broker. The Indian wedding broker is essentially a matchmaking middleman, a consultant-for-hire with vast personal databases of eligible men and women, sorted on a range of parameters from caste and family money to looks and career prospects. Money matters most of all; the broker's first question before taking on a task tends to be "How much can you afford?" Like real estate agents, brokers receive a small percentage of the total monies spent on a wedding. The best brokers can make millions in a year, matching the sons and daughters of India's richest families.

The Aroras' broker turned out to be a dud.

"My parents introduced me to sheer morons!" Simran exclaimed, as we sat in the garden lounge of a five-star hotel in New Delhi. We had picked a hotel café near her home because her parents had long imposed an 8:00 p.m. curfew on her, as was typical of many Delhi families with their daughters (but not their sons). "No kidding. I was rejected everywhere," she recounted, able to laugh about it now that her luck had changed. "One guy rejected me because he didn't like my parents' house. Another guy told the broker I didn't laugh enough. Like, hello, who says that, dude?"

The introductions were held at either side's home, closely chaperoned by both sets of parents. The broker would often be in attendance. Over the course of a year, Simran was introduced to more than thirty men. At first, as she said, she was game. "This is how it's done," she thought. A year later, it had begun to feel like torture. The introductions were almost always stilted, with the broker's presence lending a transactional air to the proceedings. Simran's ego took a battering: none of the suitors came back with a declared interest. Several of the men seemed emotionally stunted, not knowing how to start small talk with someone of the opposite sex. One suitor, on seeing Simran dressed in a traditional salwar kameez, inquired, "Can I see you in Western clothes?" before asking, to be clear, if she wouldn't mind stepping out and changing into jeans right away. Another prospective groom, probably in search of long weekends away, asked to see if she could drive. (Simran refused to oblige in either case.)

"Apart from one guy, everyone rejected me," she said.

And what about that one guy?

"He was shorter than me. Sorry, that's my one nonnegotiable. And smokers. No smokers, please. Can't do it."

The rejections contradicted Simran's merits. Physically she was an attractive individual, a little taller than average, with a symmetrical face, large, searching eyes, and long eyelashes. Her skin was pale and fair, which, given Indians' disdain for darker complexions, must be seen as an advantage. Simran worked in the human resources department of one of Delhi's top media companies. She spoke in a casual mix of Punjabi/Hindi and English—"Hinglish," as it's known—but in an Americanized way with liberal sprinklings of "like," "dude" (as above), and "no kidding." She had a self-deprecating and winning sense of humor: "I recruit men for a living, but I can't get one for myself."

Simran gradually began to lose faith in her parents' ability to set her up. The Aroras, it turned out, had a small, closed circle. The broker's database was little better—he didn't seem to understand who Simran was or what she wanted. None of the families they knew were compatible with the person Simran had become: outspoken, confident, independent. Men seemed to want pliant, younger wives who would look pretty and stay at home. That wasn't Simran. The arranged system had met its limits.

Meanwhile, Simran was nearing twenty-eight, which in Delhi's Punjabi community is perceived to be too old to marry. All the suitable men were taken, as the thinking went.

Simran decided to broaden the search. Her parents were not well versed in the internet, so she guided them into the world of online matchups. "I told them the internet was like a virtual broker," she recalled. "Same thing, but without the fee."

On a friend's suggestion, the Aroras turned to Shaadi.com. This seemed like a reassuring option because the Hindi word *shaadi* means wedding; the website was no place for mere dalliances. The Aroras had only recently invested in a PC and internet connection at home, and so every day, father, mother, and daughter would sit around the PC and pore over profile after profile. Income? OK, will do. Hindu? Tick. Punjabi? Tick. Engineer? Excellent! Father? Doctor—yes, respectable family. Star sign?

While Simran was hoping to amplify her search, the Aroras were encouraged by Shaadi's ability to do the opposite: one could filter by a choice of "mother tongue," religion, caste, and community.

The results were unsatisfying. Every day, dozens of email matches would pop up in Simran's in-box, which she would then show to her parents. "I think I rejected thousands of matches in total," she said. "Many of the guys on these matrimonial sites were very direct. One of them said, 'I want to do a webcam with you.' Are they insane? What kind of site do they think they're on?" she added, lamenting how uncivil the men she encountered tended to be.

Months passed by. The Aroras were becoming desperate. Why couldn't they find a suitable boy? What was wrong with their daughter? What was wrong with the stars?

Simran tried other matrimonial sites. Sometimes standards would be lowered and numbers would be exchanged. Text or WhatsApp conversations would start with the basic questions. Favorite movie? Song? Food? But she rarely met with anyone. On a couple of occasions, family meetings were arranged, but they were strained affairs. The parents were uncomfortable meeting someone they hadn't vetted through a broker or friends. Simran found it awkward to orchestrate; she was unable to chat privately with a single match.

Something had to change. Simran decided to try smartphone dating apps. Perhaps she could indeed fall in love? Tinder had just arrived in India. It was a simple concept. The app would filter prospective dates within a given radius. Swipe left on a picture for no, swipe right for yes. If both sides voted yes, the app would introduce the willing couple, who could then meet up right away for anything from a coffee to sex. "There's a stigma with these apps—they're seen as 'hookup' platforms," admitted Simran. "But there are also a lot of serious guys looking to get married. I had to give it a try."

The Aroras were a conservative Hindu family, so Simran knew a casual romance was out of the question. In any case, given her family's 8:00 p.m. curfew, anything more serious than an early evening drink was difficult to arrange. But she remained optimistic that she'd have luck finding someone on her smartphone. "The main thing is that phones offer us something we never had growing up here: P-R-I-V-A-C-Y." She spelled it out for emphasis, explaining how her parents would crowd around her while she surfed the family PC. "Now I can search privately on my phone without my parents commenting on the guys, and I can meet them on my own without being surrounded by a million family and friends. Even if we don't actually meet, we can text, dude. Otherwise how do you get to *know* a guy in India?"

Simran began meeting men on Tinder. She kept it secret from her parents. There were more than a dozen dates around Delhi: men who were rich, poor, cute, ugly. Most had very little experience dating. Simran had been to an all-girls school and college in India. Even in England, she couldn't bring herself to open up to men in her yearlong master's program. "We don't really have a culture of doing that. We're so coy. It's such a novel concept to go to your parents and say, 'Hey, I found someone!'"

Dating apps gave Simran choice, but with that came the fear of failure. "If my dad picks someone for me, and it doesn't work out, he'll just say, 'Hey, deal with it, that's the way it is.' And I will. But if I meet someone on a dating app and things don't work out, then my parents will blow up—it'll be about how I'm doing it all wrong, how our generation has lost its Indian culture. So the stakes are higher with this thing," she said, holding up her phone. "We have to be careful because we're doing something edgy, you know? We just have to get it right."

Around the time Simran still believed her parents would find her a match, the Delhi-based entrepreneur Sachin Bhatia was looking at ways to enter the matchmaking business. Bhatia had spent a decade working with MakeMyTrip.com, a pioneering web start-up. As more Indians moved to the cities to work, there was new money to be spent on leisure. MakeMyTrip became a popular way for them to research and book getaways.

In 2010, when MakeMyTrip went public, Bhatia quit in search of a new calling. Once again, he homed in on a growing national trend: love. Young Indians were looking for options other than being set up by their families, but, having had little exposure to the opposite sex, many didn't know where to begin with dating. India already had several matrimonial websites, but nothing involving casual romance. In Bhatia's mind, the key to this was the smartphone, which allowed for the privacy of looking, swiping, chatting, flirting, without parents stepping in. Bhatia already knew about Tinder and Grindr in the West, but dating apps that focused primarily on location—essentially a tool for a quick hookup—were in his opinion a step too far for the average Indian. There was a gap in the market for helping millennials discover dating.

Bhatia was poring over possible ideas with two entrepreneur partners: Rahul Kumar, a former colleague from MakeMyTrip, and Hitesh Dhingra, a founder of Letsbuy. One day, while they were brainstorming at a New Delhi café, a song playing in the background seemed to strike them like a thunderbolt. It was "Truly Madly Deeply," a 1997 song by the Australian pop band Savage Garden. (Pop hits from the 1980s and '90s have tended to have a longer life in India than in other parts of the world.)

The lyrics:

> I'll be your dream, I'll be your wish, I'll be your fantasy
> I'll be your hope, I'll be your love, be everything that you need
> I'll love you more with every breath, truly madly deeply do...

It was a eureka moment for Bhatia. *Truly, Madly, Deeply*. Truly, to create an app based on truth and trust, with verified profiles of real people; Madly, to do something unconventional, cool, out of the box; Deeply, to form a lasting connection with someone.

And so Truly Madly was formed.

In a low-trust society like India, the founders knew, there would be immediate skepticism as to whether profiles were real. Truly Madly attempts to counteract that with a robust system of checks and balances.

Getting on the platform requires several forms of identification: driver's license, college ID, connecting to one's Facebook or LinkedIn account. Truly Madly conducts its own checks as well by calling the applicant's family and workplace. (A quarter of all prospective users are rejected. If one is discovered to be married, for example, there is an instant expulsion—affairs are mad, even deep, but not true.)

Trust is a crucial factor when it comes to dating and marriage. In the traditional matrimonial column in Indian newspapers, for example, ads are often placed by the parents of prospective brides and grooms. Families have developed a strange and unique code of abbreviations in these ads. SM4 is "suitable match for"; grooms are described as h'som, brides as b'ful; PQ means "professionally qualified," and BHP stands for "biodata, horoscope, and photo." Since the process is controlled by two sets of parents, there is an element of vetting—and therefore a building of trust—before a meeting is arranged.

Truly Madly's system of verified social capital seeks to replace the traditional role played by families and brokers. The difference is in the detail. While matrimonial ads highlight caste and community, Truly Madly showcases the authenticity of a profile through pictures and lists of one's favorite movies and songs, imported from Facebook. Prospective dates can add in their education and income levels, which for young Indians seems a better predictor of a match than caste. The more detail, the more sources of information, the greater one's Truly Madly social capital. After that, the app is very much like Tinder: swipe left for no, swipe right for yes.

"Women prefer men with higher trust scores on the app," Bhatia told me, as we chatted in the open sitting area of Truly Madly's headquarters in Delhi one afternoon. The office space was previously an art warehouse. The large rooms were cheerful and airy, with walls painted in bright yellows and pinks. A few billboards of the app's advertising campaigns hung on the walls.

When it first launched, in 2014, Truly Madly made a splash with what it called a #BreakingStereotypes campaign. Twenty-something men and women were each pictured holding a declarative signpost. "I'm Fair and I don't care if it's Lovely"—a reference to India's $500 million fairness-cream industry, led by Unilever's well-known Fair and Lovely brand. Or "I'm Hindu and I heart beef burgers," "I love shopping but my dad is not my ATM," and "I wear red lipstick and I'm not easy." The point being: one could be a *real Indian* and still go on dates.

"It's a tough sell," Bhatia admitted, explaining that despite rave reviews for the app's design and safety features, there was still a lingering social stigma. "Because of the taboos about online dating, a girl will not tell her best friend that she met a guy on Truly Madly."

In most cases, Truly Madly matches don't actually meet in person. But that doesn't make it futile. After a couple is matched, they migrate to the app's chatroom to suss each other out. The average user spends forty-two minutes on the app every day—more than on Facebook—chatting, sending "stickers," flirting. This is where Indians are learning the mating game. It is their safe space, without the judgments of their elders. It is also a place that offers an escape. According to Bhatia, young women in small towns across India were using Truly Madly to target men in the big cities. "A lot of our early adopters were girls from small towns who had moved to Delhi, Mumbai, or Bengaluru," Bhatia said. "Truly Madly was their window to the world." The young women would then tell their friends back home about smartphone dating. So girls in villages and small towns—with no chance of having the freedom to date locally—would log on to look at profiles of men in the cities. And then the chatting and texting would begin. "It might be taboo, but dating is still aspirational," Bhatia said. "That's why there is an untapped market."

Nearly half of Truly Madly's users are between eighteen and twenty-two—about college age. Eighty percent of the app's users are men. "Men are desperate," said Bhatia. "Women play it safe." It is slim pickings for the men, given the short supply of women. On Truly Madly, for example, if a man gets a single "like" back from a woman, without things going any further at all, the man will probably stay on the app for another seven months. "Love is a very hopeful thing," he said.

Bhatia's observation—that Indian men are desperate in their search for love—is frequently borne out in stories in the news. In a February 2017 investigation, the *Hindustan Times* journalist Snigdha Poonam uncovered how mobile recharge stores—where people go to add prepaid voice and data plans to their phones—were involved in a nefarious harassment racket.[2] The shopkeepers would save the numbers of their female customers, ranking the women according to their looks; then they would sell those numbers to men willing to pay for them. According to Poonam, a store in Shahjahanpur in Uttar Pradesh was selling numbers of someone considered "beautiful" for as much as 500 rupees, or $7.50. The number of an "ordinary looking girl" would sell for a tenth of that amount. The men would call these girls, harassing them to meet up.

Surprisingly, sometimes cold-calling a woman's number can lead to success. While researching female entrepreneurship in Bangladesh, Juli Q. Huang of the University of Edinburgh stumbled upon what she has begun to call "wrong-number relationships."[3] Huang was spending time with mostly young, unmarried, poor women in rural areas. That's when she noticed something different. "A call would come in, a random number," she told me. "The women would pick up and say hello. The other person would say hello, and then they would ask, 'Who are you?' I would have gotten irritated. But these women would actually respond. They'd strike up a conversation." Huang found that only a small percentage of these chats would turn romantic, but she realized that there was a large group of men randomly dialing numbers to chat with people. "The calls were always initiated by men . . . but for women, it was giving them an experience—of chatting with a man—that really wasn't possible for them before," she said. "It revealed a lot about society and loneliness in that part of Bangladesh."

If the supply of potential dates is seemingly low, so is the level of skill at seduction, says Bhatia. "We found out that guys sometimes didn't know what to say on our messenger app. They didn't know how to say 'Hello, you have a nice smile' or 'You have a beautiful name,'" he pointed out. "Indian men aren't very good at making small talk."

Bhatia decided to create TM Spark, a paid service that allows users to send messages to other profiles even if they haven't been matched. Truly Madly experts guide users in what to say by looking at their Facebook profiles and information on record. "So, let's say, if both of you like Coldplay, we'll tell you to strike up a conversation about the band," said Bhatia. "And if we don't find something in common, the app gives you standard icebreakers on favorite movies or authors." Bhatia calls these prompts "smarter starters"—there is a higher probability of women answering positively to them than to a regular conversation.

Bhatia thinks the smartphone will end up transforming romance for young Indians. "This kind of dating would not have been possible without the phone," he said. "If you're sitting at the doctor's, or waiting for a taxi, you browse Truly Madly and it draws you in. The minute you get a 'like,' it's a major validation. You get a thrill. And all of this happens on the go in a way that couldn't have happened in India before. For most Indians, dating wasn't accessible without the phone.

"But we're still very Indian," added Bhatia quickly. "Our USP is that we are exclusively for singles—not married people looking to cheat. Just because you know a girl is a kilometer away doesn't mean you can have her," he said, alluding to the notion that dating apps had a reputation for quick GPS-enabled hookups. "Location doesn't matter so much here. It's like a Bollywood movie: you have to win your partner over."

Simran was seated in the coffee shop of the five-star Leela Palace Hotel in New Delhi. Sitting across from her was a young woman in a sky-blue top and jeans. She was Honey Chopra, a well-known astrologer in the city. She had a laptop open in front of her. Simran had sought Chopra out to ask the one question on her mind.

"Am I ever going to get married?" she blurted out. "Because if I'm not, then just tell me now. *I'm exhausted! I keep meeting the wrong men!"*

Chopra was peering into her laptop screen, where she had compiled Simran's horoscope using the exact date, time, and location of her birth. "Yes, there is a man in your destiny," she pronounced, a coy smile slowly appearing on her face.

"What? Really? What is he like?"

Chopra, continuing to interpret from the chart in front of her, began to describe what she envisioned. The man Simran would marry would be tall... broad-shouldered... have wavy hair...

Simran liked what she was hearing.

Chopra continued with a string of descriptions, continuing to stare into her computer: ambitious... knowledgeable... mentally stable... not religious but well versed in the scriptures...

Simran mulled this over. Chopra seemed to be waiting for a pointed question.

"Will he be smarter than I am?"

"He will be your superior," came the response.

This pleased Simran. She wanted to be with someone she could admire and look up to. That would be an improvement over the people she was meeting so far.

When would they meet? How?

To this, Chopra could not divine a precise answer. "Soon," she replied, uncertain but reassuring. "Soon."

When Simran recounted to me her meetings with Chopra, she had a sheepish look on her face. Astrology was not in vogue among Delhi's young professionals. But Simran was a believer. "Look, it's a science just like any other science," she said. "And there are so many instances in my life when astrology has been proven right."

According to Chopra, an astrological chart reveals a road map for an individual's life. "It's just like GPS, showing you which way a road is headed. It's a blueprint for your future," she told me.

Chopra is part astrologer, part therapist, with the ability to listen to her clients' problems for long hours, interjecting suggestions every now and then. She has an almost hypnotic, lyrical voice, and she often disappears behind her laptop screen as she interprets the charts in front of her. For Chopra, astrology is an unlikely career because her training was in recruiting, with a degree in advanced management from Cambridge University in the United Kingdom. But when she studied astrology and started predicting things accurately, she was hooked. "It was kind of creepy how many things I was getting right," she told me. Now astrology is her full-time job. "A lot of my clients would be uncomfortable admitting they follow astrology," she said. "But it is the basis of a lot of their decisions."

Hokum or real science, there is no doubt that astrology is big business in India. By some accounts, Indians spend more than $10 billion annually on consulting astrologers. Like most other Indian industries, astrology could be moving to a smartphone-based model.

"The market for online astrology is already nearing a billion dollars," said a bullish Vaibhav Magon, the CEO and founder of an Ahmedabad-based start-up called AskMonk.

Magon's company has a simple model: to connect young Indians to prevetted astrologers on its platform. The customer can ask questions, pick an astrologer from a list of options, make a digital payment, and then wait for a response. The astrologer will consult the customer's information—the time and location of birth—and divine the future. The two never need to meet.

"The old way of consultations was cumbersome," said Magon. "You need to first find someone you trust. Then you have to make an appointment, travel a distance to meet them in person. Sometimes you have to go with the whole family, which can be weird. We cut all of those things out."

AskMonk promises responses within twenty-four hours. Eighty percent of its users are between the ages of eighteen and twenty-nine. The most common question, predictably, is "When will I get married?"

I've always fallen into the camp that sees astrology as dubious, and I think of it as a tradition that will eventually die out in India. Magon, like Chopra, disagrees. "Astrology has been around for four thousand years!" he told me. "More than 80 percent of all Indians have been to an astrologer at least once. This is here to stay."

I question why young Indians like Simran—progressive, modern, and increasingly Westernized—will turn to astrology to divine answers about their decidedly New Age problems. Magon thinks the opposite. "Indians will modernize, but the core beliefs never change. Astrology will always be sacred here."

Perhaps it was written in the stars that Simran would turn to Truly Madly. Everything else had failed. With the practiced delete-finger of a recruiter, she had rejected thousands of profiles on matrimonial sites. On Tinder, the right guys didn't seem to swipe right. Not only that, she felt ashamed to be seen on the app. She worried a colleague would see her, or worse, a cousin or a family member; if that happened Simran would be branded as what Indians call "fast"—too Western. Why not try something less risqué?

Simran signed up for Truly Madly through her Facebook account. Profiles of dozens of potential matches were generated. She began looking. Swipe left, swipe left, swipe left. Nothing interesting. It was a Sunday, and Simran had just had an argument with her parents. Anger was building up over their collectively futile search for Mr. Right. "What would people say?" her father asked. "Why are we failing?" Simran had no answers. She was busy pouring her frustration into Truly Madly, swiping left on male profiles with slash after slash of her index finger. "Reject! Reject! REJECT!" It was therapeutic. It boosted her ego.

She stopped at the picture of a man with wavy hair and a toothy grin. She didn't know why she stopped. She swiped right. Ooops. Could she take it back? Too late. Was it a mistake? She looked at a few more profiles, swiping left with less relish than before, deep in thought.

A few minutes later, he swiped right on her too. It was a match. They started messaging.

"Hey," she texted. "Let's connect?" Simran's tone on Truly Madly's messaging app had taken on an almost businesslike air. Dating was a serious affair, after all. There was no time to waste.

"Hi," he responded. His name was Ritesh Singhania. (As with Simran, I've changed Ritesh's name. He stands to get into even more trouble than Simran if his parents find out what happened.)

"Here's my number. Want to talk on the phone?" Simran didn't like texting. As a recruiter, she found it hard to gauge a person's qualities in the casual syntax of text-speak. But she felt that when she heard someone's voice, she could instantly size him up. The accent, the tone, the timbre—it contained all the markers she needed.

Ritesh called right away.

His voice was strong, firm, his English to Simran's liking. "How's your Sunday?"

"I could kill for a drink," came Simran's response. She surprised herself with her own boldness. It must have been because she was still stewing from arguing with her parents.

"I'll come and pick you up. Address?"

And with that, Simran had set up what would be the most important Sunday evening in her life.

An hour later, a dark green sedan pulled up in the lane outside Simran's home. It was 5:00 p.m. She slipped out of her parents' house and walked toward Ritesh's car. The German-made vehicle was a good omen in Simran's mind: her date was clearly well-off. She knocked on the window and sat down beside the driver. She barely looked at Ritesh as he started the engine. Where would she like to go? "Hard Rock Café," she declared. She had never been, but she felt like a drink and rock music.

Simran turned to look at her date. She smiled as she took him in. He had a kind face. He was a little plump but well built. Simran approved of the preppy red polo and faded khakis. They began chatting. He worked at a bank. He went to college in the city of Pune. They were the same age. Why Truly Madly? He was just trying it out, he said; this was his first time.

Simran noted that her date wasn't a creep. His eyes weren't roaming all over her body. He was a gentleman.

The new couple's first disagreement wasn't far away. When they found themselves a booth at the Hard Rock Café, they discovered what is usually a nonnegotiable mismatch in India. She was a vegetarian; he liked chicken. She wanted a drink; he was a teetotaler. Uh-oh.

Simran considered the downsides and chose to ignore them. She liked Ritesh's company so far; he didn't seem to care.

The next three hours flew by, as Ritesh and Simran chatted excitedly about their lives, their stories, their hobbies. Much of those three hours was a blur. But in the background there was Linkin Park, Pink Floyd, Bon Jovi, margaritas (for her) and chicken nuggets (for him.)

It was nearing her 8:00 p.m. curfew. Ritesh offered to take her home. "Which guy doesn't try to break curfew?" she thought at the time, pleased that Ritesh seemed a man of moral fiber, willing to accept her parents' rules—even though they were both touching thirty.

There was no kiss goodbye. That wouldn't be right. With a wave of her hand and a smile, Simran said good night. Ritesh honked the car in response. But the date didn't end there. Back at their respective homes, they began messaging on WhatsApp. Work stories. Travel. Family. The clock ticked on. They argued about politics. They teased each other. Their lives revealed themselves, slowly painting a picture of who they were. School lives. Previous girlfriends (many). Past boyfriends (one). His ideal woman (slim). Her ideal man (rich). It was as if they yearned to know more, to talk more, to connect more deeply.

Dawn broke. Neither had slept. Adrenaline, excitement, and curiosity had sustained them all night. Simran had to see him again. She asked where he lived. It turned out he was her neighbor. She ran out of the house, telling her parents she was going for a jog. It was 6:00 a.m. There he was, just a hundred meters down the road, standing right by a parked dark green German car. She jogged up and stopped a foot away, beaming. Ritesh smiled back, bleary-eyed.

Simran was beginning to believe in love.

Three heady months passed. Ritesh worked nearly fifteen hours a day, and Simran couldn't stay out late, so they sneaked in lunches near his office a few times a week. They were getting to know more about each other, beginning to form the makings of a relationship.

They quickly realized they had two problems hanging over them. Simran and Ritesh both believed in astrology. What if their horoscopes didn't match? This, of course, would immediately halt their budding romance. And second, if they told their parents about their relationship, they couldn't say they met through a dating app. Their parents would disapprove. Simran and Ritesh needed to conjure up a lie.

But first they had to decide whether to get their astrological charts matched. Simran told me about their dilemma. The irony was clear: on the one hand, they were defying convention by shaping their own destiny, but on the other hand, they were believers in fate and astrology, just like their forebears. Simran confronted Ritesh about what they should do. It was almost as if they were waiting for the results of a biopsy.

"Let's do it before we get too close," said Ritesh. It was a cold, practical answer. Simran was of a similar mindset. She went back to see Honey Chopra again, armed this time with the time and place of Ritesh's birth.

"So, is it going to work?" asked Simran anxiously.

Chopra was deep in thought.

"What's wrong?" cried Simran. "Is there a problem?"

"It's a match, but . . ." Chopra's voice trailed off. Simran was already leaping with joy. And then: "You will marry him, but you're never going to get on with your in-laws. Ritesh's chart shows he has a deeply troubled relationship with his parents."

Simran was relieved and dismayed all at once. How could an Indian marriage succeed if the bride didn't get on with her in-laws?

From a distance, the sounds of a marching band, first softly, then loudly, steadily building up to a crescendo. At the head of the procession was a young man on a white horse. The horse was decorated in colorful sequins and glitter. On top, the man—the groom—wore a tight-fitting white sherwani, a formal suit with a long coat, elaborately brocaded with gold. He wore a red turban coiled tightly around his wavy hair, perspiring despite the February cool. He was beaming, gently bopping as the drums and the trumpets blared out a medley of Bollywood tunes.

Around the groom were his friends, six or seven of them, dancing awkwardly around the horse as it trotted forward. One of the dancers was doing the Macarena. Another was hopping on one foot as he clapped his hands. As they inched forward, the traffic on the roads was halted, forced to wait for the procession to pass. Car horns added to the cacophony of the trumpets.

The groom's procession was to arrive at the wedding venue to be welcomed by the bride's family, the hosts. Then the ceremonies would begin.

So far, so typical. Indian weddings are large, loud, garish spectacles. For rich and poor alike, a marriage is an event that calls for extravagant amounts of buying, catering, feeding, and spending. Until I left India, at the age of eighteen, the only weddings I had been to were Indian ones, with hundreds of guests invited and much singing and dancing. But when I moved to the United States and made my first American friends, I began to realize how different the two sets of cultures were—and therefore how unique the weddings of my childhood were. When the first of my college friends got married, in Boston, I attended and marveled at the contrasts. At the time, I jotted down some observations. Here they are, with some elaborations.

Differences between Hindu weddings in India and Christian weddings in America:

1. American weddings generally have anywhere from thirty to a couple of hundred people in attendance.

   Indian weddings tend to have hundreds of people in attendance, sometimes flowing into thousands.
2. In the United States, there is no guarantee that you will be invited to a friend's wedding. In fact, it is a real honor, a prize that means you are a *Very Close Friend.* The bond is further ranked and solidified if you make the cut as a bridesmaid or groomsman—then

you're really part of the inner coterie. Dinners are meticulously planned and seated. Speeches are made.

In India it is the opposite: you or your parents invite everyone you possibly can, from your dentist to your doorman (this, despite India's affinity for hierarchy). Friends can often bring friends; the more the merrier. A wedding is a free-for-all. Dinners are almost always buffet, not seated.

3. In the American wedding, there is a clear moment when you know that the couple has gotten married. It happens when, at the end of the vows, the priest says, "You may now kiss the bride." Everyone cheers and throws flowers. They're married.

   In India, one rarely gets to know when the couple is actually, formally, wedded. Often the ceremony takes several hours, with hundreds of guests milling about in the background, chatting and eating various courses of food, ducking in and out as they show their faces at other weddings nearby as well. (Since weddings tend to take place on auspicious days in the Hindu calendar, they are clustered with other weddings.) Richer families throw parties that are so big that sometimes they have overflow banquet halls with the ceremonies broadcast live on large screens for guests to watch along.

4. The American wedding is clearly about the couple. It is their moment, their coming-out party for all of society to see. The bride and groom have a first dance to their favorite song; they pick the band and the playlist. They often pick everything from the food to the décor and the flowers.

   The Indian wedding tends to be the opposite—it is about everyone *but* the couple, whose choices and preferences take a backseat in the face of large families organizing multiday soirees, catering to countless family traditions and religious ceremonies. Indian weddings are unions of two individuals, but more than that, they are unions of two clans.

Generalizations all, of course. The point is that Indian weddings are unlike almost anything else. They are chaotic affairs. Most of all, they include large numbers of even the most distant family members. And that's what was missing at the February wedding I was attending. As the groom's procession arrived at the four-star hotel that was the venue, and I recognized the Aroras queuing up with their aunts and uncles and cousins to welcome their son-in-law-to-be, it struck me that the unthinkable had indeed happened. Ritesh's parents weren't there. In fact, none of his family could be seen, only a handful of friends and work colleagues. Of the hundred-odd people present at the wedding—which was small by Indian standards—most of the guests were Simran's family.

I didn't dare ask what had happened at the wedding itself. Not knowing any of the guests, I stood back and observed. Simran put on a brave face, but I could tell she was troubled. Ritesh seemed tense. The traditional Hindu ceremony, with the couple walking around a fire seven times, was conducted in the middle of the hotel's reception hall. As tradition demanded, the bride's parents were seated near Simran. But where Ritesh's parents should have sat were two men who looked like people from his office.

A couple of weeks later, I caught up with Simran. She had just come back from her honeymoon. At first, she was reluctant to talk about Ritesh's family, but it didn't take long for her to open up. "We did everything the right way. We took their permission. I just can't believe this is how it turned out," she said, clearly still in shock.

She recounted what happened. In the days following Chopra's verdict, the happy couple set about telling their parents. They concocted a story about how they met at a work event in Delhi. Simran's media company was doing some business with Ritesh's bank, they said. They were introduced, liked each other, and started dating. It was a mild diversion from the truth. Neither Ritesh nor Simran wanted to lie to their parents, but they felt they needed to win their approval—and telling them they met through a dating app would have undermined that. "They've given birth to us," said Simran. "You really have to make sure they are on board with your choices."

When they learned of the budding romance, the parents on both sides wanted to do things the traditional way. They wanted "bio-datas" of their prospective in-laws—a one-page document with family bios and photographs. Simran and Ritesh both obliged. That's where the first hint of problems began, said Simran. She was a Punjabi; Ritesh was from Rajasthan's Rajput community. Simran's family were vegetarians; Ritesh's relished meat. Like most Punjabis, the Aroras liked to drink their scotch; the Singhanias preferred chai. Suddenly, the odds seemed almost generationally, historically, culturally stacked against them.

Then, on seeing Simran's photograph, Ritesh's parents found something troubling. They told their son they needed to see a thalassemia test. "They told Ritesh Punjabis are vulnerable to thalassemia, and so they wanted to make sure I wasn't affected," said Simran, describing to me how the blood disorder worked. "It felt insulting. I did it anyway. And I was negative. But the test made me wonder—are they trying to find reasons not to like me?"

Simran and Ritesh loved Bollywood movies and were deeply influenced by the 2014 hit *2 States*. As the title suggests, a young Punjabi boy falls in love with a female Tamil classmate at business school. The two date for a couple of years before they finally tell their parents—on graduation day. Chaos ensues. Neither set of parents is happy about their child marrying someone from a different culture. But Punjabi-boy and Tamil-girl persevere, going to comical lengths to win over their prospective in-laws and bring the families together. Along the way, the boy fixes a troubled relationship with his father. Simran and Ritesh felt that the similarities with their own lives were eerie. They figured the best course of action was to follow the movie's plot: to try to smooth over any parental differences.

Simran and Ritesh tried to orchestrate a family meeting. Problems quickly emerged. Simran's parents wanted to invite the Singhanias home, since they were traveling from the nearby city of Jaipur. The Singhanias, however, wanted to meet the Aroras at a neutral venue—at a Delhi hotel café.

A compromise was reached. The Singhanias would first meet Simran at a hotel café. Then they would all go together to the Aroras' home.

Simran was nervous. Her first meeting with Ritesh's parents was a blur. "His mother seemed very fashionable," she recalled. "I somehow got through it all."

The Singhanias arrived at the Aroras' home with a large suitcase in their car. When they opened it up, they explained how they had come prepared: they had brought sweets, incense sticks, and prayer mats. They had come to perform the traditional Hindu ceremony to declare an engagement. "In Ritesh's caste, it's called a Roka," explained Simran. *Roka*, which means "to stop," signifies that a young woman and man are now taken off the market; they are officially betrothed.

"I was freaking out," said Simran. "I mean, we knew we wanted to get married and all—eventually—but none of us knew his parents were going to do the ceremony the first time they met my parents! It's not like we could say no. It all just happened so quickly. And then Ritesh's mom whipped out her phone and started Skyping her brothers and sisters. The next thing I know I'm chatting with his entire family. I was so overwhelmed!"

Simran remembered being happy despite it all. "We were in love. They made us sit together as they performed the engagement ceremony. My parents were crying in happiness. And right at that moment, I realized, sitting next to Ritesh, that our shoulders were touching as we sat together. That was actually the most we had ever touched so far."

Just like that, Simran and Ritesh were engaged. "And that's where the trouble began," said Simran. "You want to know a fun fact? His parents are just effing weird. They wanted to ruin this all along."

A week after the impromptu engagement ceremony, Ritesh's mother began to take charge. She wanted Simran to wear *mehendi*, the patterned henna designs that Hindu brides-to-be often wear on their hands. Simran didn't take the suggestion seriously. A friend painted a comical circle in henna on her hand. Ritesh's family wasn't amused. Other requests were made: wear a saree; send us pictures of yourself in a bridal gown. Ritesh's parents wanted to send pictures of their pretty daughter-in-law-to-be to their friends.

"So weird," Simran recounted to me. "I mean, who makes those kinds of requests anymore? And I don't even remember how to wear a saree. Punjabi girls don't really wear sarees the way they do in Rajasthan," she moaned. "Dude, they were looking for ego massaging."

Ritesh's parents complained to their son. "Why is she disrespecting us? See? We told you. She's not from our culture!"

Relations slid downhill. Soon Ritesh's parents stopped speaking to the couple; they refused to be involved in the wedding plans; they emotionally blackmailed Ritesh to give her up.

"What pisses me off is that we did the right things throughout. We didn't elope. We sought our parents' approval. We got their approval. *They* got us engaged. And then . . . " Simran's voice trailed off. "And then they tried to mess it all up."

Honey Chopra says this was all too predictable. "I had told them that Ritesh's mother was going to pick fights with Simran," she said.

"Simran was overreacting too—it's in her charts. She could have done more. But the real problem was Ritesh. He was never close to his parents. The Singhanias were in a strange, almost generational power struggle, and Simran got stuck in the middle of it."

The rest was a game of chicken. The Singhanias expected their son to accede to their wishes and call off the wedding. The bride and groom believed the Singhanias would come around at the last minute—just the way it happens in movies like *2 States*.

"All they needed to do was show up," said Simran. "When they didn't, Ritesh told me quite bluntly, 'They're dead to me.' And we went ahead with the wedding as planned."

A couple of months later, I asked to meet Simran once again, this time with Ritesh, whom I had spent very little time getting to know. I wanted to understand how things could have gotten so bad with his parents—couldn't he convince them to get on board? There was also the question of smartphone dating. I wondered whether introducing something as radical as the phone had disturbed the centuries-old harmony of Indian marriage.

When the three of us sat down for a coffee, I tried to break the ice with a game. Before they met each other, I asked them, what were the top qualities they had been looking for in a prospective partner? Their answers somewhat surprised me.

Ritesh

1. "The horoscopes have to match. If they don't match, no marriage."
2. "My wife shouldn't be working. When I first told Simran this, she didn't like it, but this is the way I feel it should be." Simran was quiet as Ritesh described his view of things. "My criteria are simple: you see me off in the morning, and when I come home, you should already be home. In between those hours, you can do whatever you want." Simran nodded silently, then said she didn't mind this—she only needed to work for six or seven hours a day, and because Ritesh worked long hours, she was able to meet his stipulations.
3. "My partner needs to be a homemaker who can take care of my life and my house—stuff like electricity bills and mobile bills. I can't manage it. I need a house manager. I have fifty pairs of shirts

and fifty pairs of socks because I don't know how to do laundry. That was very important to me." Again, I didn't detect an adverse reaction from Simran. For the first time, the loud, bold Simran I had first met a year ago seemed more timid and tamed—and not unhappily so.

4. "She needs to respect my family." As he said this, Simran and Ritesh locked eyes and stayed silent.

Now it was Simran's turn.

Simran

1. "No smoking."
2. "A guy should have to have the ability to build a life. He has to be resilient in the face of setbacks."
3. "He needs to always be loyal and honest with me."
4. "He has to be superior to me." As Simran said those last words, it struck me that she and Ritesh had established a clear hierarchy in their relationship. She had accepted him as a superior, as a greater intellect and mind, as someone whose needs would be catered to above hers.

I was confused. What about shared values? Love? Humor? Wit? What about a mutual interest in books, or movies, or the theater? What about travel? What about the future, and children, and money, and savings, and friends? Simran and Ritesh both laughed at me as if I were being silly. "No, we just discover our lives together. We don't worry about this stuff!" they both said, almost in unison.

I tried to switch topics and asked what really happened with Ritesh's parents.

"I don't know," replied Ritesh, sighing. It was clear he was still struggling to talk about this. "When I first told them about her, they were so excited. They saw hope that their thirty-plus son is finally going to get married. But they had unrealistic expectations. On the one hand, they found her too homely and domestic. They wanted a daughter-in-law who would look glamorous in the lobby of a five-star hotel," he said. I shot a glance at Simran to see if she was taking offense at this. She didn't seem to be. Ritesh continued: "But at the same time, my parents couldn't cope with a daughter-in-law who could speak back and had a mind of her own. My parents are well educated and modern when it comes to everything...apart from when it comes to treating a daughter-in-law. That's the problem with our society."

It turned out Ritesh had a history of dating girls and then being unable to get his parents' approval. There was the girl in high school who would pop by to go over homework with him. When Ritesh's parents found out there was something more going on, they threw a fit. Ritesh was forced to break up with the girl. The same thing ended up happening with at least three girls from his college days, he said.

"It became a pattern, actually," Ritesh told me. "Something was always wrong. Either the family wasn't well-off, or they came from the wrong caste, or they didn't have the right house or car, or the horoscopes didn't match. There was always something. I used to go into depression, and I thought I would never get married."

What I didn't understand from chatting with Ritesh was how the Singhanias had been able to dictate his life to such an extent. "I don't know," admitted Ritesh. "Financially, socially, mentally, I was completely dependent on them. I was a mama's boy. We were high-caste Rajasthanis; we had honor and pride. And I wanted to make them happy. I wanted to do the right thing and make sure I was taking my parents' advice on the most important decision of my life."

After Ritesh moved to Delhi, the more independent he got, the less he cared about his parents' views. "I bought my first phone in 2009. They didn't like me spending money on expensive things. Then I bought my first car in 2013. That was a real act of rebellion. They were savers. I'm from a different generation, a different mindset. I believe in spending and enjoying. I want a luxury SUV. I want holidays. I want a fine life. They don't understand these things."

I asked Ritesh why he thought his parents would reject Simran. How did they go from being so excited about her that they set up an impromptu engagement to taking the nuclear option and skipping their only son's wedding? Such an act, after all, was almost unheard of in India.

Ritesh described how he flew to Jaipur several times in the months leading up to the wedding. He tried to convince his parents to change their minds. But their conversations would always end with bouts of screaming and emotional blackmail. "They didn't respect me," said

Ritesh. "I had to yell at my father, *'I'm thirty-one! Everyone in my company looks up to me and respects me!'* Why can't he accept I've grown up? I still don't understand. We tried and we tried. It was a standoff. They thought I'd just back down like I always did. I didn't. And they lost."

Simran doesn't feel like she won. "I wish his parents had been there to witness that day. That day was magical for us, but it will never come back for them. They just missed out on what should have been the happiest day of their life."

I asked Simran and Ritesh if they would have found each other if it weren't for a smartphone dating app.

"No way," said Ritesh. "India has changed. In the old days, it would take weeks and months to set up a meeting between families, and it would be formal and stifled. Today, if I have fifteen minutes of spare time at midnight I can just text someone I've met on Truly Madly. Simran and I would never have met if it weren't for technology. We have no friends in common; we don't frequent the same places. We're not the same caste or community. And yet we lived around the corner from each other, and we fell in love."

Simran agreed. "In India we like to filter our prospective partners by caste, community, income. Even though we've become more modern, we still like to filter for certain things. I wanted to marry someone who was successful. Nothing else was working for me. But I met the love of my life on a smartphone. I swiped and he swiped…and I can't stop thanking God for it."

Simran and Ritesh had only been married a few months, but I could detect a rare level of openness and comfort between them. They had been through a lot. They were more practical and pragmatic than I might have expected, yet I had little doubt they were very much in love. Their circumstances—of the groom's parents not attending the wedding—were certainly extreme, but I knew from speaking with several other couples on Truly Madly that the process of winning parental approval was taxing. And, while Ritesh and Simran came from upper-middle-class families, it was clear to me that smartphone dating apps would soon reach the middle classes in a bigger way; it was inevitable.

I asked the couple if they would try to reconcile with Ritesh's parents one day.

Ritesh snorted. "No chance. I'm not even going to try."

Simran, however, said, "You know, we might be modern, we might have met on a dating app, but our Indian roots don't go away. It's ingrained in us for centuries. I am still hoping for his parents to come and bless our new home and my husband."

We said goodbye for the last time. I wished the couple well. As they walked away, I noticed their shoulders gently pressing against each other. But their hands and fingers weren't linking up: they were clasping their smartphones.

5

Cyber Sutra

The Internet Is for Porn

IN THE YEAR 2012, a generation ago in digital technology, the person who generated the most internet searches in India was not a cricketer or a Bollywood star. Nor was it a politician or a religious figure. None of them were close. The person most Indians were curious about that year—as measured by the total number of Google searches[1]—was Canadian-Indian Karenjit Kaur Vohra, a.k.a. Sunny Leone, a former porn star and *Penthouse* Pet of the Year.

It wasn't the case only in 2012. As hundreds of millions of Indians continued to discover the internet through 2013, 2014, 2015, 2016, and even 2017, Sunny Leone remained the most-searched-for person in India. People simply couldn't get enough. (Prime Minister Narendra Modi made it to number two in 2014, the year he was elected, but Leone remained the clear favorite.)

Prudish, conservative, family-values India... and a porn star? Leone was no longer even performing; she had stopped around 2010 and started her own production company with her husband and manager, Daniel Weber. In 2011, she came to India as a guest on the reality TV show *Bigg Boss*, a local version of the *Big Brother* franchise. Leone's appearance was predictably controversial (by design, of course: it was good for the ratings). Although most Indians hadn't heard of her, it didn't take long for word to spread: "*A porn star—from America—here in India?*" At the time, parliamentarian Anurag Thakur complained to the Ministry of Information and Broadcasting, arguing that Leone's presence on a nationally telecast program would "have a negative impact on the mindset of children." Thakur added: "When children see these porn stars on TV and then do a Google search, it shows a vulgar site. It will have a bad impact in the long run."[2]

There were no laws, however, to stop Leone from appearing on TV. While the production of pornography was officially illegal in India, Leone could justifiably argue she was no longer involved in the industry. She was trying to pivot to general entertainment. And in that sense, she was already on the path to stardom: her stint on *Bigg Boss* brought her the first trappings of national fame. The 2011 season launch, of which she was a part, remains the highest rated in the show's history. Seeing her on *Bigg Boss,* Bollywood director Mahesh Bhatt offered Leone a role in one of his films. Soon a spate of mainstream media opportunities presented themselves. The 2014 music video "Baby Doll," for example, became a national hit with more than 120 million views on YouTube. (The lyrics were saltier than the usual Indian fare: *Yo soniyo* [lovely], *she put up a show / Unless I'm impressed, baby I gotta go / I won't mind tell you that I take you floor / But tonight no bites 'cause the wife will know.*)

Leone's video played within the rules of Indian entertainment. She could pose sexy but not unclothed. Indian censors would never allow a movie with nudity to air in mainstream cinema (and Leone was keen to transition to mainstream stardom). But Anurag Thakur turned out to be right on one thing: India couldn't police the web. A simple search for Leone invariably brought up thousands of links to her old adult videos. Leone's own website—SunnyLeone.com, one of the first results on Google—sold a subscription to her X-rated videos and photographs for $9.99 per month. One didn't need to pay, however. Just as YouTube is revolutionizing the global consumption of video, the world of pornography has websites like YouPorn, RedTube, and XTube, each of which is owned by a larger umbrella empire known as Pornhub. (According to the tracking agency Alexa, Pornhub.com was the world's 38th-most-visited site in 2017—well ahead of the 104th-ranked BBC.co.uk and the 127th ranked NYTimes.com).

Pornhub releases an annual report showing exactly how its users consume pornography.[3] The treasure trove of data has been called the Kinsey Report of our era, a real-time look at how people are thinking about sex.[4] The data are full of quirky cultural insights, sorted by country, search terms, and kinky desires. The numbers are mindboggling: there were 28.5 billion visits to the network in 2017, with 110 gigabytes of data transferred every second. Over the course of the year, 595,482 hours of video were uploaded, which would have taken one person sixty-eight years to view, uninterrupted.

Pornhub's 2017 report revealed India to be a crucial market.[5] It was the world's third-biggest consumer of pornography, behind only the United States and the United Kingdom. (Given that India's internet user base is projected to more than double within a decade, it's fair to expect India will soon become the world's number one consumer of adult videos online.) Somewhat surprisingly perhaps, 30 percent of visitors from India were female, marking the fourth-highest proportion in the world, behind only the Philippines (36 percent), Brazil, and South Africa. Eighty-six percent of India's traffic came from mobile users—the highest on record—bearing out the theory that most viewers of porn were watching it on smaller screens and in private. The average age of a Pornhub user was thirty-five; in India the average fell to just thirty, matching its younger demographics.

There was more. Indians seemed to stop looking at porn on religious holidays: according to the 2016 report, there was a 17 percent drop in traffic from India on the Hindu holiday of Diwali, and a 15 percent drop the day Ramadan began.

Here is Pornhub's 2016 entry on India: "Most countries take great pride in their nationality, but India takes it to a whole other level as made evident by their top searches on Pornhub. The majority of searches (top, relative, and gaining) include 'Indian.'"

The top searches:

1. Indian
2. Indian wife
3. Indian college
4. Indian bhabhi (*bhabhi*, older brother's wife)
5. Indian bhabhi devar (*devar*, husband's younger brother)
6. Indian actress
7. Indian teacher
8. Indian aunty

Indian viewers of pornography weren't seeking out the high-production-value HD videos that were made in Los Angeles. They wanted to watch mostly amateur Indian videos—badly shot and produced but perhaps more authentic.

The other clear winner from India was Sunny Leone. According to the Pornhub report, Leone was the most-watched porn star among viewers from India even though she hadn't made a new adult video in years. The back catalog of her work was there for all to see, easily accessible, searchable, and free. Ever since she became known to Indians in 2011, Leone had been the most-watched and most-searched-for performer on the Pornhub network. As more Indians were getting online, predominantly on smartphones, they were discovering modern-day pornography by the millions. Sunny Leone was often the gateway drug to this new world of sex.

In a sense, Indians were vicariously fulfilling their sexual curiosities and fantasies through pornography. A top-ranking Indian telecom executive, who insisted on remaining unnamed, told me that more than half his network's bandwidth was being consumed by users downloading and streaming adult content (and evidently, a fair bit of it was of Sunny Leone). "This is India's dirty secret. Everyone's watching this stuff on the internet, but no one will admit to it. We are known as a conservative society, but the minute Indians are online, most of them seem to turn to pornography."

There were seemingly no boundaries to where porn could be consumed. In 2016, after Google installed a free Wi-Fi service at the city of Patna's railway station, it emerged that the location was a top national source for visits to pornographic websites.[6] In 2012, three Indian politicians in the state of Kerala were forced to step down after they were caught on camera watching adult videos on their phones during a session of Parliament.[7]

Meanwhile, Leone herself continued to transition into a mainstream public figure in India. She was the face of movies, started a line of erotic fiction, endorsed products, and performed at people's weddings. In August of 2017, for example, when Leone visited the city of Kochi to inaugurate a phone store, the actress herself was stunned by how many fans came to see her. "My car literally in a sea of love in Kochi, Kerala," she tweeted to her 2.4 million followers. Alongside the tweet, Leone shared a video of herself surrounded by what must have been nearly ten thousand fans screaming out, "We love Sunny, we love Sunny."[8]

It was an unlikely outcome. Even in her hometown of Sarnia in Canada, Leone had struggled to find acceptance, as she described in a visit to her childhood home in the 2016 documentary *Mostly Sunny.* "The Indian community in Sarnia doesn't want anything to do with me because of my choices in life," she is heard to say, after a scene shows an older Indian woman refusing to talk about Leone. "One-point-two billion people in India can get past it, but the people in Sarnia cannot." Leone went on to describe how India had accepted her with open arms. "[People in Sarnia] want nothing to do with me, but all of India would *die* to be a part of my family."

Leone was indeed popular in India, but in fact she didn't have an easy ride there either. In 2015, a Mumbai housewife reportedly filed a police case against Leone "for distribution of obscene content" on the internet.[9] In 2016, another police case was filed against her and other actors for a scene from the movie *Mastizaade,* in which the actors discussed condoms while they were in a Hindu temple.[10]

Perhaps the most difficult moment for Leone came during a national TV interview in January 2016. She was doing the rounds to promote a new movie. On the CNN News18 talk show *The Hot Seat,*[11] the journalist Bhupendra Chaubey introduced the actress as "contentious, controversial," before going on to declare she was "completely antithetical to what we perceive as an Indian woman."

As Chaubey began the interview, he appeared to be trying to turn the program into a confessional. He was dressed in his usual Indian formal wear: a white kurta shirt, with an unbuttoned blue Nehru jacket on top. Leone was wearing a sheer flesh-colored dress, embroidered with flowers. As the two faced each other on camera, it seemed the stage had been set: Chaubey was the pious Indian and Leone the racy outsider.

"Tell me one thing that you regret," began Chaubey. "One thing you believe went wrong for you."

Leone responded by saying she regretted not being able to get home in time when her mother passed away. Sensing that Chaubey was looking for a different answer, she went on to say she didn't have any professional regrets. "I've made mistakes," she said, "but I learned from my wrong decisions."

Chaubey seized on this excitedly. "What kind of mistakes?"

Leone described how she struggled to adapt to Bollywood. "It's a different world," she said. "There's so much chaos... I get frazzled."

Chaubey sensed his moment to shift toward the reason he was interviewing Leone. "Do you get upset? You know, there's so much written about you. I hope you know you're the most-Googled person in India. You have overtaken even the prime minister of India. So there's something about Sunny Leone that generates a lot of curiosity," he said, a smirk on his face.

Leone didn't take the bait.

Chaubey then went on to point out that famous Bollywood actors were reluctant to work with her. "There is a lot of resistance... inhibitions... about you. How do you deal with that?"

"I don't care about anyone else's inhibitions," responded Leone.

"But will your past... your past as a porn queen... continue to haunt you?" pressed Chaubey. He had finally said it: the *P*-word, five whole minutes into the interview. "Will it continue to hold you back?"

"I'm not held back," snapped Leone, beginning to sound defensive.

"But is your past, literally and figuratively, in every way, a thing of the past?" pushed Chaubey again. "I began this interview asking you if there was one thing you regret, and you spoke about your mother. But let me say that if I was to turn the clock back, would you still do what you did?"

"One hundred percent," responded Leone. (This was roughly in line with what she had told previous interviewers, that she had enjoyed her career as a porn star and was thankful for it but was now looking to transition into being a mainstream actress.)

This wasn't the answer Chaubey seemed to be expecting. "So you would still do the kind of shoots, the kind of work that you used to do, before you came here..." His voice trailed off.

"Everything that I have done in my life has led me into this seat," said Leone. "So it's a chain reaction that happens. Everything is a stepping-stone to something bigger or better."

"Pardon me if I'm being in any way offensive," Chaubey persisted. "How many people would think in terms of growing up to be a porn star?"

"No one," said Leone firmly. The slightest glint of a tear seemed to appear in her eye, but she composed herself.

"And yet you became [a porn star]," pronounced Chaubey.

A while later, there was this: "You know, there's a lot of chatter, people say, 'You see a Sunny Leone film, you'll get morally corrupted.' I'm wondering whether I'm being morally corrupted because I'm, well, I'm interviewing you."

"Well, I can leave if you want," responded Sunny.

"No, not at all, not at all," came the response.

The interview lit up social media that night. On Twitter and Facebook, users called it "insulting" and "demeaning." The Bollywood star Aamir Khan weighed in too. "I felt bad when I saw the interview, and the types of questions," he said in a press conference that week.[12] "She's a human being, she's a woman, and I respect her. I'd be happy to work with her."

The tables seemed to have turned. If the great Aamir Khan respected her, then so could India.

When I called to set up a meeting with Sunny Leone, her publicists seemed reluctant. I could imagine why. They didn't want another TV interview like *The Hot Seat*. I told her team I wasn't interested in passing judgment; I simply wanted to learn Leone's thoughts on India's smartphone internet boom and its impacts on pornography. Her team was intrigued. We arranged to meet at a Mumbai hotel.

I arrived an hour early. We were meeting at the JW Marriott near Mumbai's Juhu Beach. Leone lived nearby. In preparation, I had asked around a bit: What would it be like to interview Leone? Journalist friends and others in the entertainment industry told me to expect someone who was warm, smart, and kind.

Leone's PR team had cordoned off a private sitting area on the lobby floor of the hotel. She arrived on time with two others: her publicist and her husband, Daniel Weber, who played guitar in a rock band and acted as her manager (and had performed alongside her in some of her porn videos). The group was dressed casually. Leone, in a black T-shirt and jeans, was wearing very little makeup; I had agreed to interview her without cameras. Even without dressing up, Leone struck me as glamorous and beautiful, a head-turner in a city of celebrities.

As we sat down, Leone and her publicist had their eyes on a nearby TV screen. It was an advertisement. Leone was giggling. She was watching herself on the screen, playfully writhing on the sand by a beach. The commercial was for Manforce condoms, one of India's top sellers. Leone was the group's main pinup, selling calendars under the banner "Enjoy a whole year's worth of Sunny days!"

We sat down to talk. I began by asking about her movie *One Night Stand*, which was just out at the time. As the title suggests, it is about a married man who gets drunk and ends up in bed with another woman, played by Sunny Leone. Months later, the protagonist learns that Leone's character is his new boss's wife. But in a twist to standard Bollywood fare, no romance ensues. Instead, the protagonist spirals and his marriage dissolves, while Leone's character tries to salvage hers.

"You know, it's always OK for the guy to be a womanizer, but a woman can't want a man. It's so one-sided," Leone told me. "And that's not just in Indian cinema, it's everywhere. That's what we wanted to address and tackle."

We chatted about her role for a while before I asked Leone about the young Indian consumer—a target audience that she seemed to understand better than most other entertainers.

"Young Indians want to be successful. That's what drives them. If you have an idea or a dream here, it is actually possible. If you stay on it and you have integrity, you have the ability to do whatever you want here. I'm walking proof of that. It's the Indian Dream."

The phrase suggested that social mobility was there for anyone who strived and worked hard. The question was whether smartphones or technology played a role in its realization.

"India is changing every second," she said. "With the internet, suddenly everything is accessible here. Everything is at your fingertips in a way that really wasn't possible before. So much good stuff is available to young people. Thank the Lord that this country is a democracy and they haven't censored the internet here."

Ah, but they have, I reminded her. In 2015, India's Ministry of Communications issued an order banning 857 pornographic sites. The decree was in response to a petition filed by an intellectual property rights lawyer in the country's Supreme Court.[13] Reflecting views undoubtedly held by many in India, the petition read: "Nothing can more efficiently destroy a person, fizzle their mind, evaporate their future, eliminate their potential, or destroy society like pornography.... It is worse than Hitler, worse than AIDS, cancer or any other epidemic. It is more catastrophic than nuclear holocaust, and it must be stopped." The prohibition turned out to be short-lived, in part because of a backlash from free speech advocates, but also because the government realized it was impossible to prevent new sites from popping up and circumventing the ban.

Even so, New Delhi had a history of trying to censor the web. As far back as 2009, the government tried to ban the popular X-rated comic *Savita Bhabhi*. The protagonist, Savita, was the stuff of teen fantasy: a shapely housewife who was only too happy to have rip-roaring sex with men who weren't her husband. Each issue would have Savita in outlandish pornographic trysts, with the neighborhood boys playing cricket, her doctor, her house help, her bra salesman, or in any other scenarios one could conjure up. In the early days of the internet in India, new issues of *Savita Bhabhi* would instantly go viral.

Leone was familiar with the bans and had had her own brushes with the Indian authorities. "Visiting the police stations here is scary," she said, referring to when that housewife had filed a case of public indecency against her. "People protested, burnt my pictures, they said I was corrupting the youth or something like that. I mean, what does that even mean? I didn't invent sex, you know? But I still feel blessed to be here. I thought I was going to be in India for just two weeks, to film *Bigg Boss*. And then all these amazing things came to me."

Leone began to open up. She became chatty and confident. I guess that her accent sounded Canadian, with a hint of California thrown in—explained by her time in the adult entertainment industry in Los Angeles. Her pronunciation of Hindi words was, however, impeccable. She had been studying Hindi to prepare for roles in Indian cinema.

I had to ask her about pornography and whether some of it was damaging, especially for children.

"I think it's every parent's responsibility to put blockers on their child's phone. You need to monitor what your children are seeing. I do not want kids watching porn," she said firmly.

But what if the parents aren't tech-savvy, or even literate? Many children are the first users of that kind of technology in India, particularly among the middle and lower-middle classes.

"Look, you don't need to be literate or have a degree to have basic parental common sense," she said. "That's universal. Those feelings of protectiveness come naturally. And parents have to learn."

Leone could sense I didn't find her answer satisfactory. In my mind, there was no way a poor family in rural India would know how to install blocks on their smartphones. If Phoolwati's nieces and nephews started accessing pornography in rural Rajasthan, she would likely not know how to monitor it.

"I have trouble thinking the typical viewer is not a smart intellectual person," said Leone. "Our consumers are smart. That's why we have so many films like *Dangal* that have a beautiful, intelligent message."

Leone had shifted the argument to a more general one about quality content in India. *Dangal* was 2016's Bollywood superhit, with the protagonist—played by Aamir Khan—training his three daughters to become successful professional wrestlers. I didn't disagree with Leone's argument—Indian viewers indeed deserved more credit, and better content—but that was very different from access to pornography. I pressed the issue again. Hadn't easy access to smartphones and the internet in India opened Pandora's Box, in a sense?

"Maybe. Sunny Leone wouldn't exist without the internet, I guess. I call myself a product of social media," she said. "That's where everything started for me, with the phone. The ability to click photos, to take videos, to share them, to connect with people—that's what makes the world go round. People like to think porn is something dirty, but it can be sexy too, and empowering."

Our time was up. Leone had another appointment. She had been charming, warm, and friendly—as promised—but I was left disappointed with her insights on the impact of porn or smartphones in India. I assumed it was because she simply didn't want to talk about pornography. She had moved on. Sunny Leone was a brand, a business—and a hugely successful one at that. Even now, there were millions of Indians who were watching her simply as an actress, with little idea about her previous career. Leone was looking to the future.

The question remains, however, whether India is equipped to deal with a smartphone-driven explosion in the viewing of pornography. In *Mostly Sunny*, the documentary about Leone, a former police commissioner of Delhi, Kiran Bedi, said that increased access to porn was responsible for a higher incidence of rape in India. In the documentary, Leone's response to that was indignant: "Rape has been going on for centuries. It didn't get created when I got created.... It's just now in your face because of the internet, because of TV, because of people reporting it. It's not my fault. Stop freakin' blaming me!"

Leone was correct, I thought, in pointing out that rape is not a new phenomenon. Yet the belief that porn plays some part in the widespread problem of rape is persistent. In the aforementioned 2015 petition calling for a ban on pornographic websites, for example, the word "rape" or "rapist" was mentioned seventeen times, and the petitioner contended: "Many studies reveal that rapists and child molesters use pornography both immediately prior to their crimes and during the actual assault." The argument is not limited to India. In the United States, in response to similar theories, the Northwestern University law professor Anthony D'Amato published a paper in 2006 titled "Porn Up, Rape Down," arguing that there was in fact a correlation between a *fall* in rates of rape and increased access to pornography.[14] D'Amato presented data highlighting how the four US states with the least internet access—Arkansas, Kentucky, Minnesota, and West Virginia—reported a combined 53 percent jump in rates of rape between 1980 and 2000. In contrast, the states with the most internet access—Alaska, Colorado, New Jersey, and Washington—saw a 27 percent reduction in the rate of rape. D'Amato interpreted the data to posit that viewing pornography was a form of release, making it less likely for one to resort to violence. "Internet porn has thoroughly de-mystified sex," wrote D'Amato. "Times have changed so much that some high school teachers of sex education are beginning to show porn movies to their students in order to depict techniques of satisfactory intercourse."

The Indian sexologist Dr. Prakash Kothari agrees with D'Amato's findings. "There is simply no connection between rape and viewing pornography," he told me. "Rape comes from a mental pathology. Pornography has no role either way."

Kothari is seventy-seven and worked at one time with the late American gynecologist Dr. William Masters, the subject of the 2013 TV series *Masters of Sex*. Kothari has practiced in Mumbai for more than forty years and seen an estimated fifty thousand patients. "Films can also enhance sexual desire in loving relationships," he said. "If people have a basic understanding of sex education, then they can use pornography for their enjoyment. If not, it is to their detriment."

India's problem may be a lack of sex education, says Ira Trivedi, the author of *India in Love*. "Porn is becoming more and more violent," she told me. "Young Indians aren't sensitized to these kinds of images the way a young American teen might be. In the West, women can largely wear what they want, and boys see and understand that from a young age. In India, we have been more repressed. Our perception of women and people from lower castes is inherently a power-based dynamic. So you can imagine, whatever the negative impacts of pornography in the West, it is likely worse here. We simply don't know the extent of it."

Ranjana Kumari of India's Centre for Social Research (CSR) says that while there are no studies to point to a link between viewing pornography and rape, she is inclined to agree with Trivedi's view. "Of late, the rape cases we are learning about seem more gruesome, more heinous than what we used to hear of before," said Kumari, referring to news stories of violent gang rapes and of sexual assaults on girls as young as seven or eight. "Young boys from the villages are moving to the cities to work. They often live together in cramped rooms with little to do. With their smartphones, they end up gaining access to the most extreme, violent forms of pornography. You have to realize, these are uneducated village boys who have not seen a female body in full form. They have no supervision. They're learning about basic biology and sex through violent pornography. It's normalizing the abnormal. That's got to have an impact."

As part of her work with CSR, Kumari often travels to rural parts of India and meets with communities to discuss their problems. One thing that has begun to worry her is the recent prevalence of "rape videos"—real-life rapes that are filmed and then shared on social media. "It's so horrible it is unthinkable," she said, describing how some of the videos would gain mass distribution through WhatsApp. "It has become very common. And if it is being encouraged in some way by people watching pornography, then we have to find a way to curb it. All of this is made possible only by smartphones and the internet."

While adult content of all forms has always existed in India, access has historically been limited. *Playboy* magazine, for example, has long been available in India. When Prime Minister Jawarharlal Nehru was interviewed by *Playboy* in 1963 there was a nationwide rush to procure copies. But it remained too expensive for the average Indian. Knock-offs like *Debonair* were cheaper but, like *Playboy*, mostly limited to an English-speaking elite. Kothari recalls rudimentary magazines in Indian languages, but those were also limited in their distribution. Either way, it was hardly easy for an Indian reader to comfortably access adult content. "Young Indians don't usually have any personal viewing space," said Trivedi. "Our conception of privacy is very different from the American notion of privacy. Even if a teenager here gets his hands on an adult magazine like *Playboy*, where does he keep it? He'll get caught instantly. But with content on a phone, you can hide it, save it, and watch it again and again with no one really knowing."

Adult films also have a history in India, but again, with limited access. "Blue films," with rudimentary depictions of sex, were often screened in Indian cinema halls in the 1970s. In the 1980s in Tamil Nadu and Kerala, "bit cinema" had a moment: theater owners would insert bits of pornography in their movies to titillate hungry viewers. "But all of that was pretty much for a tiny segment of the moviegoing population," said Trivedi. Sex in modern India was mostly hush-hush. The barrier was access. At corner stores across India, one could access porn as long as one had a laptop to store it. In Mumbai, the words *dal gosht*—a lamb curry in a lentil sauce—were code for a CD or flash drive full of local porn, uploaded (or side-loaded) onto a computer for a small fee. But laptops were expensive. "In the 1990s and 2000s, very few Indians had laptops or personal computers. Even today, the average Indian college student doesn't have his own laptop," Trivedi pointed out. "What's changed in the last few years is access. All of a sudden, with a cheap smartphone and free Wi-Fi, Indians have a free pass to a world of porn."

One thing most experts seem to agree on is that India's past reveals a nation that once had a much healthier relationship with sex. This is best highlighted by the *Kama Sutra*, the well-known guidebook that was written more than 1,600 years ago. The poet Kalidasa's *Raghuvamsa*, written around the fifth century, described sex without any shyness. The eighth-century Assamese temple of Kamakhya Devi—the goddess of sex—is a shrine to sexual appetite and continues to attract millions of tourists every year. The temple of Khajuraho in Madhya Pradesh and the caves of Ajanta and Ellora in Maharashtra each depict sexual encounters and positions in graphic detail. Even during the Mughal Empire, a period of Muslim rule that began in the early sixteenth century and lasted for three hundred years, miniatures depicted sexual positions. "But once India began to attract foreign influences, the curtains fell," said the sexologist Kothari. "Propriety was first on the agenda. Sex came last."

In *India in Love*, Trivedi argues that British rule and Victorian mores brought to the surface an ascetic side of Indian culture, "especially when the British discovered the high-priest of prudish values—Manu." *The Laws of Manu*, a set of Sanskrit teachings from around the third century BCE, was adopted by the British as a primary Hindu text, propagating "many negative assumptions about lower castes and women that sharply restricted their freedom, regulated their behaviour, and blocked their access to social and political power."

Twenty-first-century India seems once again to be shifting to a new direction, toward more openness. "Urbanization and access to technology is pushing against the Victorian-Indian mindset," Trivedi told me. "We are now witnessing the first major generation of young Indian women who are moving from small towns to the big cities to study and find jobs. This is already having an impact on how Indian society deals with love, sex, and relationships."

Kothari points out that the sooner India returns to an era of confronting sex openly, the better. "I've had patients come to me saying they believe they can cure a sexually transmitted infection by sleeping with a virgin. That's the state of Indian sex ed. Even our medical colleges don't have courses on sexuality. With smartphones, the era of cyber *Kama Sutra* is already upon us, and in a mass way. Why keep living in the past?"

6

India's iGen

Growing Up with Nomophobia

SOMETIME IN THE MIDDLE of May in 2017—at the height of summer in India—a grainy mobile phone video began to make its way across the country. The recording rocketed from phone to phone, first in the eastern state of Jharkhand and then nationally, circulating among groups on WhatsApp.

The video was shot in portrait, showing a man beaten and bloodied, crumpled up on a patch of barren earth. His white undershirt was rolled up to his chest and drenched in blood. Encircling him was a small mob of men armed with cane sticks. Several appeared to be filming the proceedings on their phones.

"You son of a bitch," someone screamed. "Motherfucker! We'll kill you!" A cacophony of abuse was under way. The man was pleading for his life, but his cries were drowned out by a rising tide of expletives and fury. The mob continued to beat the man.

The video cut to black.

The subject, Sheikh Haleem, was killed. He was only twenty-eight. Six others were killed as well, across two separate vigilante attacks. It was as if a cloud of rage had suddenly descended on a small part of Jharkhand, propelling village men to embark on extrajudicial murder sprees.

It turned out that a rumor had spread that a group of strangers was abducting children from nearby villages. The rumor made its way onto WhatsApp; the rumor morphed into "news"; the news, circulating from phone to phone, villager to villager, was weaponized; a group of locals decided to act. The rest happened very quickly. A mob was formed. Strangers were produced, beaten up, and murdered. Justice was delivered. The recordings of the killings were duly sent back out into the ether of WhatsApp, completing the cycle of horror.

Jharkhand's police were befuddled. There had been no reported cases of child abductions. The rumor was completely unfounded.

"Rumors have always flourished in India," says Pratik Sinha, the founder of a myth-busting website, AltNews.in. "But it's become exponentially dangerous because of the internet."

Sinha is a former software engineer who decided in 2017 to dedicate his life to fighting what we've come to describe as "fake news." He collected his savings to start AltNews, which publishes daily pieces debunking fake national stories.

"Social media platforms like Facebook and WhatsApp have created silos," he told me from his home in Ahmedabad. "People read unverified things, skewed towards what they already fear or know. They believe it. And then they share it."

From my own experience, I knew this to be true—well before the constant connectivity of the internet era. As a child, I had witnessed the beguiling power of rumor firsthand when in September 1995 all of India seemed gripped by a frenzy. Word had spread that idols of the elephant-headed Hindu deity Ganesha *were drinking milk*. When a family friend told us the news in Calcutta, my mother decided there was only one thing to do: we had to see it out for ourselves. She grabbed me and my sister and rushed us to a nearby temple. The scenes were unforgettable. Long queues of men, women, and children lined up, carrying bottles and tins of milk, waiting for their turn to make an offering to Lord Ganesha. Once they reached the front, they would hold up a spoonful of milk to the idol's trunk, waiting for their offering to be accepted.

Lo and behold, the milk seemed to disappear! Grown men were crying and shaking their heads. Women hailed a miracle. There were a few skeptics—who pointed to the milk falling to the sides—but no one listened to them. When my mother's turn came, the snout seemed unwilling to suck up milk from her spoon. I tried as well. No luck. "Lord Ganesha must be full now," concluded my mother, not wanting to disappoint us.

The whole thing, of course, was a giant hoax, leading to mass national hysteria. Scientists have since explained some of the instances of milk being drawn into the clay and marble idols as capillary action. But hundreds of millions of Indians simply believed what they wanted to believe.

Sinha remembers the milk mania too. "Now imagine how quickly the rumor would have spread if people had had smartphones." Sinha does concede, however, that technology could also have helped to dispel the milk story: "It can work both ways."

The problems of rumor, superstition, and fake news—while clearly global in nature—are made worse in India's villages, where internet users often lack the savvy of their urban counterparts. While city folk generally have more experience with smartphones and the internet, rural Indian users are relative newbies. "Most villagers still have feature phones, not smartphones," confirms Sinha. "They don't even use their browsers. The web for them is limited to WhatsApp." In other words, the internet can be little more than group texting.

WhatsApp was founded by two former Yahoo! engineers who went on to sell the company to Facebook for a record $19 billion in 2014. Compared to other messaging apps, WhatsApp has a basic, no-frills interface that allows users to text, create messaging groups, and make audio and video calls on the internet. Perhaps because of its simplicity, it has an outsized audience in developing countries. Out of its 1.3 billion global users, 220 million are in India. (Twenty-eight percent of Indians, 38 percent of Indonesians, and 53 percent of Brazilians use WhatsApp.)

In a sense, WhatsApp has allowed Indians to leapfrog email. While an email address can often be a form of identity in the West, in India one's identity is tethered more closely to a mobile number. Not coincidentally, signing up for WhatsApp only requires three things: a cellular device, a mobile number, and wireless internet. You can send messages, videos, pictures—all for free.

There's a downside to leapfrogging email, however. The earliest users of the internet, mostly Westerners in the 1990s, learned and evolved with their mistakes on the web. "Chain emails" come to mind. They usually went like so: "If you don't forward this email to 7 of your closest friends you will never be lucky in love." The dire warnings would vary, but the effect was the same: mass spam. Early internet users developed a sense of email etiquette. They learned what to send contacts on email and, more importantly, what not to. Spam filters became smarter at preventing junk mail and advertising from reaching in-boxes. Crucially, users got smart about what to click on and what to avoid, which claims to believe and which to be skeptical of.

India's netizens are somewhat behind the curve. For most new internet users, WhatsApp seems to replicate the chaos of the early days of spam and chain emails. (In a typical week, at least one person on one of my Indian groups has a birthday. She is wished a happy one. And then seemingly every other person on the group sends out the same "happy birthday" greeting—to the whole group of recipients. On New Year's Eve in 2016, Indians sent each other 14 billion messages on WhatsApp, including 3.1 billion images and 610 million videos.)

The real problem with WhatsApp, however, is fake news. Unlike phone calls and email, which the government is said to have an ability to tap, WhatsApp messages are secured by end-to-end encryption, making them virtually impossible to trace. Once a fake news story emerges, there's little one can do to control it. "There's actually no way of policing it," admits Sinha. "Unlike with Facebook or Twitter, there's no way of checking what's going on inside WhatsApp. One can weaponize a fake news story very quickly."

The greatest challenge, again, tends to be with users who aren't sensitized to the very notion of fake news—and those are the fastest-growing base. While India's urban users are expanding by 7 percent a year, rural users are growing at a rate of 22 percent. With the launch of Reliance's Jio 4G service, which handed out free data in its first year, the average consumer's data consumption was surging from 2 MB a month to more than 1.3 GB—an exponential leap (1 gigabyte equals 1,024 megabytes, or 2^{10} MB).

"Most Indian users are very susceptible to falling for a superstition or being fooled by a fake news story," according to Sinha. New users, who discover a rudimentary version of computing on their mobiles, have no prior experience with Photoshop, for example, or doctored videos. "There's simply no radar for detecting BS," says Sinha. "People believe everything they see."

Hundreds of fake stories have gained national notoriety—and virality—because of WhatsApp. Some have preyed on the worst instincts of a society that is displaying an increasingly Hindu-nationalist bent.

In September of 2015, Mohammad Akhlaq, a Muslim, was dragged from his house and beaten to death on suspicion of storing beef in his kitchen. It later emerged that pictures of what resembled beef and cow parts had been shared on WhatsApp. The images were fake. Another Muslim, Mohsin Sadiq Shaikh of Pune, was beaten to death by radical Hindus in 2014 after derogatory pictures of Shivaji—a medieval king popular among Maharashtrians—were shared on social media. The killers seemed to target Shaikh because his beard implied he was a Muslim. (There were no links between Shaikh and the derogatory images.)

Sinha says he spends much of his time debunking stories from right-wing propaganda sites, including Postcard.news, Hindutva.info, DainikBharat.org, TheLotPot.com, and many others. "These sites don't cite links for facts. They don't have references. They don't have reporters on the ground. They just make up stuff," Sinha told me.

Some of those sites originate political content to benefit India's ruling Bharatiya Janata Party. Recent hoaxes have included a WhatsApp message claiming that UNESCO had rated Narendra Modi the best prime minister in the world; another claimed that India's new 2,000-rupee notes had a microchip hidden inside so they could be tracked. Neither, of course, was true. The story about the microchip was picked up and reported by national news media for days until it was debunked. No one offered corrections.

Sinha is just one of a growing cohort of myth busters on India's internet. Pankaj Jain, a Mumbai-based businessman, runs SMHoaxSlayer from his home at night. (SM stands for Social Media.) He is a self-funded vigilante activist, correcting stories he sees on WhatsApp and the internet.

Like Sinha, Jain views fake news as an old phenomenon, now potentially more dangerous because of WhatsApp. He likes to recall the well-known story of how in 1938 the actor Orson Welles and his Mercury Theatre troupe read an adaptation of H. G. Wells's novel *The War of the Worlds* on the radio. Timed to mark Halloween, this narration of aliens attacking Earth was so realistic that listeners believed it, allegedly causing mass panic. "Now imagine if they had smartphones then, and everyone called and messaged each other in panic? The chaos would have been intensified."

SMHoaxSlayer was born out of a sense of frustration. When Jain's aunts and uncles would forward him WhatsApp messages—the ones that were like chain emails—he would often correct them. It was 2015. "Back then, fake news tended to be pranks. The targets were relatively soft ones," he says, recalling a WhatsApp message that claimed Apple would design its next iPhone to look like an Indian bracelet. The rumor circulated around India, driven in part by national pride. Another WhatsApp rumor, that an employee of a soft drink company had added HIV-positive blood to a popular fruit juice, also got significant traction before it was finally debunked.

"But then things started getting nastier. We began seeing fake news about politics and religion. If religious tensions broke out, videos of violence from years past would surface, purposely inflaming the mood."

Communal tensions have been particularly sensitive to internet rumors. On May 16, 2017, the *Times of Islamabad* ran a story—datelined Srinagar—claiming that the writer Arundhati Roy had insulted the Indian government in a speech. The article claimed Roy had said the Indian Army would be incapable of fighting in Kashmir even if it increased its presence there by a factor of ten.[1]

The story grew rapidly, with versions of it circulating on WhatsApp across South Asia. Several Pakistani outlets gleefully ran versions of the story. India's TV channels went one better, leading their nightly newscasts with debates as to why India's Booker Prize–winning novelist would disparage her own army. The Indian actor Paresh Rawal tweeted: "Her birth certificate must be a regret letter from [the] maternity ward." TV anchor Arnab Goswami called Roy a "one-book whiner wonder."

No one stopped to do basic journalism. A quick internet search or a phone call would have revealed that the *Times of Islamabad* wasn't a real newspaper—it was a nationalist Pakistani site. Roy hadn't been to Srinagar in months. The comments attributed to her had been made up.

"The WhatsApp messages goad you. They say, 'If you're a true patriot, forward this to ten people.' Indians react to that and forward fake news everywhere," notes Jain. "No one seems to care where the news actually comes from. There's no media literacy here. And it seems the real news media falls for these scams too. That's how it spreads."

Some hoaxes have multiple lives. One popular example is a video seeming to show a woman who is surrounded by a mob and then set on fire. The recording is explicit, showing the woman screaming, in flames, burning to death as bystanders look on. On Indian social media, the video has popped up more than once, claiming the victim is a Hindu girl burned alive in the state of Madhya Pradesh. WhatsApp forwards of the video often claim that the perpetrators are Muslim. "Again, totally false," says Jain. "This is a well-known video from Guatemala in 2014. It was mob justice for an alleged crime, but it had nothing to do with India. Anyone with common sense could have figured this out," he adds.

Neither Jain nor Sinha is comfortable volunteering where India's fake news comes from, but they concede more of it tends to benefit right-wing groups—for now. "It's Pandora's Box," says Jain of India's smartphone-driven internet boom. "And how much can people like us really stop? Our work is just a drop in an ocean of misinformation."

As a child, I was struck by a song from Roald Dahl's *Charlie and the Chocolate Factory*. It was sung by the Oompa Loompas, the colorful and slightly ominous creatures who run the place.

The most important thing we've learned,
So far as children are concerned,
Is never, NEVER, NEVER let
Them near your television set—
Or better still, just don't install
The idiotic thing at all. . . .
IT ROTS THE SENSES IN THE HEAD!
IT KILLS IMAGINATION DEAD!
IT CLOGS AND CLUTTERS UP THE MIND!
IT MAKES A CHILD SO DULL AND BLIND
HE CAN NO LONGER UNDERSTAND
A FANTASY, A FAIRYLAND!

Some would argue that you could easily substitute the word "television" with "smartphone."

One day, Saikat Sinha's fourteen-year-old eyes refused to focus on the blackboard at school. His corneas were bloodshot. His clothes were disheveled. He was sleepy but too tired to even attempt to nap on his desk. Saikat's hands were shaking. His vision was blurry. He seemed to be looking around the classroom as if through the crosshairs of a rifle, his left eye instinctively scrunched up, his right eye comically enlarged.

"Saikat!"

He snapped to attention. His biology teacher was yelling at him. Saikat took a few seconds to process what was happening around him, blinking rapidly. His fellow eighth graders were chuckling. Evidently the teacher, who was livid, had been calling his name for a while. She walked up to him, grabbed him by the ear, and dragged him to the school principal's office. Saikat was surprisingly compliant.

After much tut-tutting, a call was made to Saikat's parents. They arrived an hour later to pick him up. Saikat was sullen, angry, unrepentant. His parents didn't know it yet, but their son was clinically addicted—to his smartphone. Specifically, to smartphone games. (I have changed Saikat's name because he is a minor.)

A few months ago, Saikat had visited a cousin in the city and played *Call of Duty* on an Xbox. It was life-changing. He was hooked. Back at his hometown of Mirzapur, in West Bengal—where none of his peers had expensive Xboxes—he told his father he needed his credit card to buy some books on his smartphone. Saikat used it to purchase games like *Call of Duty: Strike Team*, *Star Wars: Galaxy of Heroes*, and *Battlefield Combat: Black Ops 2*. Mr. Sinha, a self-confessed Luddite, didn't know how to check what his son had purchased.

Saikat would stay up for most of the night, hunched over in his bed, furiously swerving left and right as he launched grenades and took aim at his digital foes. After school, he would play all the way home on the bus, stopping only for a quick lunch with his mother before escaping back to his room to play. Not knowing much about technology, his parents thought he was studying. Saikat was once a regular on the local football field nearby, but he had abandoned physical sports. The virtual world was far more immersive.

The Sinhas were only mildly concerned; Saikat's grades were still above average. But when they were called into the principal's office they realized something needed to be done.

"He is addicted," Mr. Sinha confessed to me. "Actually, he is more than addicted. He can't live without it. If we take the phone away from him for a little while, he gets visibly restless and angry. He starts panicking. Now that I know what's going on, I'm really worried about him."

After speaking to a physician, the Sinhas realized Saikat's addiction needed to be treated like any other addiction: with medical help. They were referred to a clinic in Bengaluru called SHUT—Service for Healthy Use of Technology. The clinic was run by Manoj Kumar Sharma, who was spending an increasing amount of his time researching and treating patients for smartphone addiction. There was a name for this New Age disease: nomophobia, or no-mobile phobia.

"It's a very modern anxiety that afflicts us all to some extent," Sharma told me. "I usually look at four *C*s to help me diagnose the severity of nomophobia—craving, control, coping, and compulsion."

Sharma lamented that there were few authoritative studies examining the impacts of smartphone abuse, especially in India, but from his own years of practice he could tell there were clear trends under way.

"Children don't spend enough time on the playground these days. They're not learning about normal interactions in the same way," he said. "The classic symptoms of nomophobia in children manifest themselves in being irritable and having outbursts of anger. They also tend to have a low tolerance for regular discussions, wanting to control the outcome of everything they do. Oftentimes we see signs of clinical depression."

Most of his patients tended to come from the upper-middle classes in Indian cities, he said, though he was beginning to get reports of excessive smartphone use among children in rural India as well. "Children studying in urban hostels don't want to go back to their rural homes because they don't have the same access to the internet there. Their smartphones don't catch fast data signals in the village," he explained.

The impacts of excessive screen time are likely to be the worst in rural India, where children who have access to phones are mostly unsupervised; the smartphone—as the family's first phone, computer, camera, TV—can be an overwhelming new Swiss Army knife of a toy. And parents, anxious to give their children the very best of life's opportunities, often don't know how to set rules for smartphone use. "What worries me is not what we already know about this phenomenon, but how much we don't know," Sharma says. "Imagine the damage technology is doing to children's brains?"

In the United States, which has a head start when it comes to technology, data answering Sharma's question are just beginning to emerge. The addictions expert Nicholas Kardaras says unfettered access to tablets and smartphones is damaging a generation of new users. In his book *Glow Kids*, Kardaras points out that glowing smartphone screens can stimulate the brain's pleasure center as much as sex does. "The brain-orgasm effect is what makes screens so addictive for adults, but even more so for children with still-developing brains that just aren't equipped to handle that level of stimulation," he writes. Brain imaging studies now conclusively show, says Kardaras, that "excessive screen exposure can neurologically damage a young person's developing brain in the same way that cocaine addiction can."

Some countries have begun to take these emerging problems seriously. China has identified "Internet Addiction Disorder" as a serious health crisis, while South Korea has opened hundreds of tech-addiction rehab facilities. India has been slower to react. Sharma's clinic in Bengaluru is perhaps the best known in the country, while a few others, like New Delhi's Uday Foundation, have informally begun to run "smartphone de-addiction" programs.

The impact of excessive smartphone use in teenagers may not show up in individual cases the way it visibly does for Saikat. But when they are combined to put together a tapestry of experiences, a terrifying picture begins to emerge. In a September 2017 essay in the *Atlantic*, the American psychologist Jean Twenge asked: "Have Smartphones Destroyed a Generation?"[2] The answer, it seems, could be yes. Twenge points out that rates of teenage depression and suicide have skyrocketed in America since 2011, when smartphones became commonplace; teens who spend more than three hours a day on electronic devices are 35 percent more likely to have a risk factor for suicide. Twenge calls today's American teens "iGen"—a generation that is less likely to date than Gen X or Gen Y, less likely to have a driver's license, and unwilling to do small errands for pocket money. The number of teens who spend time with their friends every day dropped by more than 40 percent between 2000 and 2015. iGen teens spend most of their time at home, not because they don't want to see people but because they don't need to: their social life is lived on phones.

The impacts of this isolation are clear, says Twenge. Eighth graders who spend ten or more hours a week on social media are 56 percent more likely to say they are unhappy than those who spend less time online. In sum, says Twenge, iGen teens are "on the brink of the worst mental-health crisis in decades"—and much of that can be attributed to a sharp jump in the time spent on smartphones.

For now, there are no identical studies on Indian teens, but the trends are easy to imagine. Ninety-two percent of Indians with smartphones keep their devices within direct reach at all times, according to the B2X Consumer Trends report in 2017 (the global average, by comparison, is 85 percent). The average smartphone user in India spends three hours actively accessing the internet every day, according to a 2016 study by the research agency Kantar and the Mobile Marketing Association. Indian teens often exceed adults in smartphone use.

I once asked fourteen-year-old Saikat who was his best friend at school.

Saikat looked puzzled at my question.

It almost feels as if Saikat likes his smartphone—a $300 Chinese-made OnePlus 3—more than he does people. Saikat is shy, reserved, scrawny. His eyebrows seem permanently furrowed. He is bright, brilliant at math and physics, full of questions about the world. But chatting with him about his life, I'm struck by how different it will be from mine. When I was sixteen, one thing I truly wanted was to learn how to drive. The car gave me a sense of real, physical freedom. Saikat laughs when I ask him whether he will ever learn how to drive. He doesn't think he'll even learn to ride a motorbike. I guess he'll have Uber—that is, if he cares to leave home. Will he develop an interest in sports, or will he keep playing video games? Will he learn how to flirt with girls in person, or will he instead rely solely on texting, Instagramming, and Snapchatting?

When I ask Saikat how India is changing because of the smartphone, my question is met with a raised eyebrow. Saikat doesn't know a different India. His generation is unique. Rich or poor, all his friends at school are online—they text, chat, message, send stickers and emoji, flirt, all on the virtual worlds in their palms. Saikat reminds me more of an

American teen than an Indian one—in part because his formative experience is being lived not on the Indian street but on the World Wide Web. I wonder if that will make Saikat and his peers more susceptible to the kinds of problems Twenge describes for the American iGen. I wonder how soon those problems will emerge.

Sharma believes India is heading toward a catastrophe unless a major public awareness campaign is initiated. "We have to make sure people understand how addictive technology is, and what it can do to our brains. We don't even know the extent of the implications right now, because things are changing so quickly here. If we don't stop to think about this, who knows what could happen?"

No other country has as many young teens as India. Once smartphones truly become ubiquitous, India's iGen will be 300 million strong—roughly the same as the entire population of the United States.

In an interview published in the *New York Times* in May 2017, Evan Williams, the cofounder of Twitter, made a startling observation: "I think the Internet is broken," he said. "And it's a lot more obvious to a lot of people that it's broken."[3]

It was a strange comment given that Williams was a serial starter of internet companies. He helped create Blogger during the dot-com boom, then Twitter in the mid-2000s, and later Medium, which he described as "a beautiful space for reading and writing—and little else." Somewhere along the way his feelings about the internet had turned sour.

Williams was not alone.

Sean Parker, whose early stake in Facebook made him a billionaire, admitted in a November 2017 speech that the social networking site's mission was "How do we consume as much of your time and conscious attention as possible?"[4] That mindset led to the creation of the "like" button, to encourage users to upload more content in the hope of receiving appreciation from friends.

"It's a social validation feedback loop . . . You're exploiting a vulnerability in human psychology," he told an audience of hundreds at a conference in Philadelphia. "It literally changes your relationship with society, with each other. It probably interferes with productivity in weird ways. God only knows what it's doing to our children's brains."

Another former Facebook executive, Chamath Palihapitiya, said he felt "tremendous guilt" about his work as the site's vice president for user growth. Speaking at a Stanford Business School event in November 2017, he said that the "the short-term, dopamine-driven feedback loops that we have created are destroying how society works."[5]

Even Mark Zuckerberg—the creator of Facebook—indicated misgivings about his work. In a public Facebook post on September 30, 2017, he wrote: "For the ways my work was used to divide people rather than bring us together, I ask forgiveness and I will work to do better."[6]

One reason that so many current and former tech executives were publicly criticizing their own work—or at least expressing remorse—was because of the events of 2016. It had become increasingly clear that social networking sites had been used to peddle fake news. Not only that, but fake news had specifically targeted swing voters in both the 2016 US presidential election and that year's "Brexit" vote in the United Kingdom—potentially swinging the outcome of the votes. For example, a CNN investigation into fake news stories in the lead-up to the 2016 US election traced a portion of the wrongdoing to a Macedonian riverside town called Veles.[7] Dozens of website operators there had created more than a hundred sites to produce fake news favoring Donald Trump over Hillary Clinton. Some of the headlines: "Michelle was caught cheating with Eric Holder—OBAMA IS FURIOUS!!!" "Bill Clinton loses it in interview—admits he's a murderer." "JUST IN: Sarah Palin hospitalized after being hit by car." According to CNN's Isa Soares, who reported the story, one of the fake news website operators she spoke with was earning $2,500 a day from advertising.

But fake news—while openly malicious—was clearly a product of an internet that was already broken, especially when it came to the dissemination of news. As the Columbia University professor Tim Wu has outlined in *The Attention Merchants: The Epic Scramble to Get Inside Our Heads*, new media outlets like BuzzFeed had already pioneered techniques like "headline optimization," which made clicking on pieces virtually involuntary. "In the hands of the headline doctors," wrote Wu, "a video like 'Zack Wahls Speaks About Family' became 'Two Lesbians Raised a Baby and This Is What They Got'—and earned 18 million views." What BuzzFeed founder Jonah Peretti was aiming to do was to trigger "pleasure in the social process" of sharing these pieces.

Basic human psychology shows that while human beings are comfortable with not knowing everything—quantum physics, for example—we usually hate feeling left out, or not knowing something that everyone else seems to know. It's what has come to be known as FOMO, or fear of missing out. Like Buzzfeed, the social media company Upworthy has taken this insight to new heights of experimentation. Its teams write dozens of headlines for every article, testing them online to see which are the most likely to be clicked on. The results often look like this: "What this child Hollywood star looks like today will surprise you. Pic #5 is especially shocking" or "6 out of 7 Americans get this movie trivia completely wrong." It makes you want to click.

The former editor of the *New Republic* Franklin Foer described these challenges in his book *World Without Mind.* The pursuit of audience is the central mission of sites like BuzzFeed, Vice, and the *Huffington Post,* he wrote. "They have allowed the endless feedback loop of the web to shape their editorial sensibility, to determine their editorial investments. Once a story grabs attention, the media write about the topic with repetitive fury."

One result is a decline in trust for the media in general. The polling company Gallup has been surveying Americans' trust in the mass media since 1972; in 2016, it recorded the lowest-ever levels of trust, at 32 percent, down from a peak of 72 percent in 1976.[8] A general climate of mistrust allows fake news to flourish. Worse, the information silos created first by cable TV and later by websites targeting specific audiences allow "filter bubbles"—where one-sided opinions and biased news can be read unquestioningly. The internet was once heralded as the tool that would make access to information democratic, free, and fair. In a sense, it has demonstrably begun to do the opposite. We've been played.

At a much deeper level, the internet has already changed how we think. In becoming our map and our alarm clock, our calculator and book, smartphones have been slowly eroding some of our oldest skills. In a 2008 essay in the *Atlantic* asking "Is Google Making Us Stupid?" the journalist Nicholas Carr wrote of how he felt he had lost the ability to read and digest long passages of writing because the internet had changed his behavior.[9] "Immersing myself in a book or a lengthy article used to be easy.... That's rarely the case anymore. Now my concentration often starts to drift after two or three pages. I get fidgety, lose the thread, begin looking for something else to do," Carr wrote. The reason, he surmised, was that he had been spending most of his time surfing the web; he had gotten used to the distractions of hyperlinks and pop-ups and alerts; his consumption behavior had changed, perhaps irreparably.

Carr's experience is by no means unique. Haven't we all sometimes felt the same way? If there is a way to fight back, however, perhaps the answer lies in more awareness, in understanding the tools that internet companies are using to monopolize our attention.

In mass media, the greatest thing that has changed in the last decade is how advertisers can reach us. In the early 2000s, companies would vie for our attention with advertisements on billboards, in magazines and newspapers, on TV shows, and on our computer screens. But we had a form of escaping from these ads; we couldn't carry our TVs with us everywhere, after all, and we would usually discard magazines after a quick perusal.

"Smartphones changed the advertising game completely," wrote Roger McNamee in a January 2018 essay in the *Washington Monthly*.[10] McNamee was an early investor in Facebook but has recently been calling for regulating the power of large technology companies. "Facebook, Google, and other social media platforms make their money from advertising. As with all ad-supported businesses, that means advertisers are the true customers, while audience members are the product," he explained. It is a terrifying thought; we, the people, are unwitting products being packaged, indexed, and sold to advertisers. I have often been struck by how I can browse the internet for, say, a pair of jeans. The next day, advertisements on my browser will keep reminding me that the jeans are still waiting to be purchased. Helpful, perhaps, but also one step closer to an invasion of privacy. And this is true not only of the West but also of India and the entire world. We are now living in an era where algorithms can determine exactly what kind of news, entertainment, and advertising to show us, based on our browsing history, age, gender, and other markers. We are being fed content by a machine that is getting paid for feeding us. India—and the world—could easily be filled with millions of Saikat Sinhas in the coming years, gorging unfettered on things they don't really need.

PART THREE

The State

7

Big Brothers

Internet Shutdowns and Internet.org

THE RAT-A-TAT OF AUTOMATIC GUNFIRE burst through the morning air in Anantnag. For the locals of this troubled district in the south of Kashmir, it came as a shock but no longer a surprise. Separatist militants had once again clashed with army forces. Several civilians were caught in the crossfire. One died; three others were wounded.

It was Saturday, July 1, 2017. Hours earlier, at midnight, India had adopted a new national sales tax, designed to stitch the country's twenty-nine states together into one economic union. The new system—known as the Goods and Services Tax, or GST—was heralded as an economic reform that would spur growth, enlarge the tax base, and make it easier to do business. Kashmir was the only state still debating whether to join. It was a symbolic outlier.

Some distance from the gunfire, sixteen-year-old Zeyan Shafiq was just waking up. He hadn't heard the shooting; his home was well insulated. When he opened his eyes, he told me, the first thing he did was to reach for his iPhone. He looked at the screen and sighed. The wireless internet at home was down. So was mobile data. Shafiq got out of bed, put on his slippers, and shuffled toward the front door, where he knew he would have a stronger mobile signal. No luck. He couldn't catch the internet. Shafiq looked up at the skies, opened his lungs, and let out a bellow of frustration.

For Shafiq, it was easy to guess what had happened. There must have been what locals called an encounter—a skirmish between Kashmiri separatists and the state. These days, encounters were inevitably followed by the government shutting down the internet. The digital blackouts weren't aimed at stopping separatists or terrorists from communicating. They were usually already dead. The shutdowns were to prevent people from sharing videos and photos of the violence on social media. In effect, 13 million Kashmiris were collateral damage, unable to do something as simple as check email.

There was a time when curfews were merely physical, imposed with barbed wire, barriers, and troops on the streets. Those still exist. Now there was an additional and perhaps more powerful one, what the Kashmiris call the e-curfew.

It turns out India is a world leader in digital blackouts. Only Iraq and Syria come close, in second and third place respectively. India's government enforced fourteen shutdowns in 2015, thirty-one in 2016, and seventy in 2017, according to the legal advocacy group Software Freedom Law Centre, which keeps a running count on the website InternetShutdowns.in. While most of the shutdowns tended to be in Kashmir, more than a dozen Indian states had been affected at various times. In West Bengal's hilly Darjeeling, for example, internet services were cut off for more than a hundred days in 2017 when a political party's demands for a breakaway state turned violent.

Meenakshi Ganguly, the South Asia director of Human Rights Watch, argues that there is a "lack of transparency and failure to explain these shutdowns," furthering the perception that the blackouts were "meant to suppress nonviolent reporting and criticism of the government."[1]

Shutdowns also have financial implications, causing steep losses to businesses. A count by the Brookings Institution estimates internet blackouts cost India $968 million in the twelve months preceding June 2016.[2] Second-place Saudi Arabia lost $465 million. Among the others filling out the top ten that year were Iraq, the Republic of Congo, Syria, and Pakistan.

If there were a ranking for individual states, Kashmir would undoubtedly have been the world's number one hotspot for e-curfews.

The result was clear: while the internet was transforming much of India, New Delhi was systemically keeping Kashmir offline.

I wondered how you shut down the internet. Is there a giant red button? How do you control which regions get curfewed? Who makes these decisions? Is there a way to circumvent the blackouts?

When I arrived in Srinagar, the capital of Kashmir, on July 1, 2017—the morning of the Anantnag "encounter"—those were the questions swirling around in my head. At Srinagar's airport, I tested my phone. I managed to catch a weak 3G signal. That morning's e-curfew was limited to the site of the shooting, some fifty kilometers to the south. I instinctively Googled "How does India shut down the internet?" About thirty seconds later, a few links began to reveal themselves. None seemed to address the question.

A local journalist, Mukhtar Ahmed, had offered to show me around the city. We had arranged to visit an ISP, or internet service provider. The company requested we keep its name anonymous.

The ISP's office occupied the entire third floor of a shopping complex in the city center. Most of the stores were shuttered: the state's trade federation had organized a protest to fight the national Goods and Services Tax. The ISP seemed to be the only establishment open that morning, presumably because we were expected. (As I discovered, businesses would frequently shut shop in Srinagar. That morning's encounter in Anantnag would turn out to be the reason for a *second* day of lost business—the next day—on July 2.)

When we entered the office doors, an orderly escorted us through to a wood-paneled cabin, where sat Mr. F, the ISP's owner.

Mr. F wore his hair slicked back and sported a bushy Mario Bros. mustache. He seemed jumpy. Before addressing me, he chatted with Mukhtar in Kashmiri. It was clear that he didn't trust journalists. After Mukhtar offered a few reassurances, Mr. F held out a limp right hand in my direction. I shook it solemnly.

He began speaking in Hindi this time, telling me about the scale of his business. Unlike the new mobile providers—Reliance's Jio, Airtel, Vodafone—he specialized in providing hardwired cable lines to large companies in the region.

"No matter the political circumstances here, the pipes are neutral," said Mr. F. "The internet is the only thing young Kashmiris trust. It is a form of freedom."

The growth of the internet in Kashmir lagged in comparison with the rest of India. I had noticed several cybercafés in the market areas in Srinagar, where Kashmiris would come to use an internet kiosk for the equivalent of a dollar an hour. These cafés—relics of the late 1990s and early 2000s—had largely disappeared from India's biggest cities. Mr. F explained that the few remaining in Srinagar were mainly for people who needed help submitting job or college applications. Most Kashmiris were now using smartphones and mobile data. His business—of providing high-speed internet lines to establishments—had also been thriving.

"But then Burhan was martyred," added Mr. F, shrugging his shoulders as if to suggest the rest was history.

Burhan Wani—or simply Burhan, as everyone called him—was a legend in these parts. He was Kashmir's Che Guevara for the digital age. Born to a middle-class family in the restive south of Kashmir, he abandoned school at fifteen and joined the Kashmiri separatist group Hizbul Mujahideen. By the time he was twenty he had become a divisional commander. But more than by his actions on the battleground—of which there is little record—Burhan made his name as a social media star. He would post videos explaining why he was fighting for Kashmiri freedom, calling on young Kashmiris to join him. Unlike other militant fighters, Burhan wouldn't bother to cover or hide his face. He spoke simply and openly, sharing his thoughts and his plans to fight the Indian Army and police. He was a handsome young man, with a Roman nose, piercing eyes, and thick eyebrows; he would often wear a T-shirt and khakis, or sometimes a kurta, with guns holstered to his belt.

In his videos, Burhan would tell Kashmiris to stay away from the police and the army, lest they get hurt during one of his group's attacks. "The only acceptable outcome is Kashmir's freedom," he would often announce. "We only have one enemy: India. The Indian state is responsible for our current situation." He also went after the Indian media's news announcements: "Kashmiris! Don't watch them—they call us terrorists, but you know what we are. We are Kashmiris like you! We want freedom like you.... If the police come to your homes looking for us, kick them out, throw stones at them when you see them coming." Burhan's videos would go viral on YouTube, the links ricocheting from Kashmiri phone to Kashmiri phone on WhatsApp and Facebook. He was inspiring—not to say inciting—a generation of young Kashmiris to fight for freedom.

Burhan was only twenty-two when he was killed in an ambush by the army. It was the eighth of July in 2016.

"Can you imagine?" reminisced Mr. F. "Tens of thousands of people showed up for his funeral. No, hundreds of thousands! They came from all over Kashmir. The police put up barriers, but nothing could stop them!"

According to media accounts from the time, an estimated two hundred thousand Kashmiris showed up at the southern town of Tral, where the funeral took place. Prayers were repeated several times to accommodate new waves of attendees. The unprecedented showing was a defiant slap in the face of the Indian government. Burhan's killing had touched a nerve; a swirl of local anger was building. What had happened to the Indian government's promises for a referendum? Enough was enough. Kashmiris didn't know where their revolution would head, but they knew one thing in their bones: they didn't feel like they were part of India.

What happened next was predictable. The Indian state cracked down with fury. Curfews were regularly enforced and with greater restrictions than before; strikes were conducted against separatists; the army attacked alleged terrorists from across the border in Pakistan. In the past, this would have been enough to break the small insurgency that had long resisted Indian rule. But this time was different. What Kashmiri militants had failed to do with guns they seemed to be achieving through social media. When protests on the streets were met with tear gas and police brutality, the images of the violence would find their way onto the web. Once there, they would spread like wildfire on WhatsApp and Facebook. The images reinforced what Kashmiris already knew: they were living in a police state. The insurgents were losing the physical battle, but they had made significant inroads into the virtual fight for Kashmiri hearts and minds.

The government realized it had to control the greatest weapon the militants now had: the internet. There was only one thing to do. Shut it down. And so began the longest e-curfew in Indian history. Month after month after month, for nearly the rest of 2016, Kashmiris had no access to the web—not on their PCs, not on smartphones.

"How does one ban the internet?" I asked Mr. F.

"It's quite simple," he replied. "We get a phone call from the police. Shut it down, or else. We comply," he said, shrugging his shoulders again. "I lost 40 percent of my business last year."

After Mr. F received the call, he would tell his chief engineer to block all data in and out, for all users apart from the ISP's main control server at Mr. F's home address. None of the company's clients would be able to use the internet. Similar moves were made by every other service provider in the state, including national-level mobile telecom companies—about a dozen in all.

"Earlier they would send us government circulars," added Mr. F. "They still do, but now those come later. It's a formality. These days we just get a call. The slightest hint of violence, and the call comes in. And if you don't comply..." His voice trailed off.

I asked Mr. F if I could look at one of the government circulars. He raised an eyebrow and looked at Mukhtar, who nodded. Mr. F relented. He pulled a large cardboard binder from one of his drawers and handed it to me.

I opened it carefully. There were dozens of stapled government circulars. I pulled out one dated June 15, 2016. The letter was addressed to "All Internet Service Licensees." The subject was clear: "Direction to block Internet Website."

> Under the powers conferred by Section 69A of the Information Technology Act, 2000, and under the Information Technology (Procedures and Safeguards for Blocking of Access of Information by Public) Rules, 2009, it has been decided to immediately block the access of the following 4 URLs.

I looked down at the URLs.

One was a link to a YouTube video. I wondered if it was of Burhan Wani. Two other links had the word "Inspire" in them, which I guessed was the name of the magazine published by the terrorist group al

Qaeda—I was familiar with it because I had once produced a TV piece about the magazine's work.

The letter went on:

> All the Internet Service licensees are accordingly directed to immediately block access to above URLs and keywords.... The compliance be submitted immediately, failing which the Department of Telecommunications may initiate action under Rule 12 of the IT (Procedures and Safeguards for Blocking of Access of Information by Public) Rules, 2009. Kindly do not reproduce the name or URLs in the compliance letter. The contents of this letter may kindly be kept confidential.

The implications of the letter were clear: citizens could roam the World Wide Web as much as they wanted, but the government had reserved itself the right to block specific websites. ISPs could do their business, but only if they did what the government asked.

"These were just a few websites. It was just the start," said Mr. F. "Once the protests over Burhan's death began, they shut the whole thing down."

And so from July 9 through nearly the end of 2016, Kashmir had no internet at all. No email, no Google, no Facebook, no WhatsApp, no friends and family, no FaceTime or Skype. No Uber, online deliveries, or maps. No research or reading or newspapers or podcasts or music. It was the ultimate modern-day punishment, imposed indiscriminately upon an entire population of 13 million. There was no recourse or ability to protest. (One government service provider, BSNL, would sometimes work—but it was very slow, and procuring a connection was difficult.)

Even when internet service was restored at the end of the year, other shutdowns awaited in 2017. Mr. F fished out another government circular, dated April of 2017. This one was an order to ban twenty-two social media sites.

"On this one they banned Facebook, Twitter, Instagram... even the Chinese and South Korean ones, all to make sure people couldn't share things on social media," Mr. F said, pointing to the mention of WeChat and KakaoTalk on the government circular.

Enterprising Kashmiris would try using VPNs—virtual private networks—to circumvent the blocks on social media sites. "But that's so slow," Mr. F lamented. "You can hardly send pictures to anyone, let alone videos. If the ban was meant to stop people from sharing details of protests and violence, then it largely succeeded. They won."

Mr. F went on to explain how these days the Kashmiri police had the top mobile operators on speed dial. They were less worried about stopping providers like him, because his reach was limited. The ones to shut down were Airtel, Vodafone, and Reliance's Jio, which had the biggest footprints. "Whoever you are, you have to comply," Mr. F chuckled darkly. "Otherwise today you lose your license, tomorrow you lose your life... who knows, eh? This is our great India."

The next morning, Mukhtar and I decided to venture inside what locals called Chhota Pakistan—"Little Pakistan." This was the area around the Jamia Masjid, the main mosque in Srinagar, built in the late fourteenth century. I was told the residents of the area were vocally anti-Indian, making it unsafe for the Indian Army to operate there.

As we approached the area, we stopped at a makeshift barrier of barbed wire, coiling in big circles across the entrance to the main street. Stores on either side of the road were shuttered, as far as the eye could see.

"We're from the press!" yelled out Mukhtar.

To my surprise, the guards didn't check our IDs. We were allowed in; they dragged aside the barbed wire to let our car through. Mukhtar turned to grin at me.

We met up with a friend of his—a photojournalist named Arshad—who lived near the mosque. His home and the entire area were under curfew.

Arshad had an insolent, cheeky smile, and he reeked of cigarette smoke. "Everyone here hates the media, but you're safe with me," he announced. "The locals remember me since I was in diapers."

We were going to visit a few victims of government violence. In the previous year, there were hundreds of incidents of police brutality. In many cases, the police fired lead pellets in the air as they chased down suspected militants. The pellets were indiscriminate killers. A single shot from a gun could release a hundred of the lead projectiles. Once in the air, they rained down like thunderbolts, tiny shards of lead striking whatever they touched. The pellets injured men, women, children, anyone in their path. In particular, they were dangerous to eyes, penetrating the retina and turning the insides into pulp.

Arshad took me to a friend's home. Mushtaq Ahmed Sofi was a baker. The entrance to his home was painted a chalky rose pink, distinguishing it from homes on either side in yellow and brown. A large window with bars revealed a one-room bakery inside, with a giant *tandoor*, or oven. Farther in was a series of smaller rooms where his family lived. We were ushered through a curtain and into what served as a small sitting room. Arshad motioned for us to sit down on a carpet on the floor.

I was introduced first to Moeen, a twelve-year-old boy. His head was tilted in a strange way. I quickly realized why: he was hiding an injury to his left eye. Moeen had lost vision in that eye, his cornea pointed aimlessly away from us. The doctors had tried everything, but they couldn't restore sight to that eye.

Moeen was a handsome young boy, with a disarming smile. He was shy at first when I asked him what happened but then explained that he had been playing cricket with his friends when the police came to hunt for militants. A shot was fired, and then suddenly pellets were everywhere. He showed me his scars. There were pitted indents on his chest, arms, back. And then there was his left eye: pulverized and rendered useless.

Moeen's mother came into the room. Like most of the other locals she disliked journalists, but Arshad had told her I was to be trusted. Her name was Sabeena. She had brought us small glasses of orange-colored soda. Sabeena was wearing a light green salwar kameez. Her eyes blazed with anger. "Why did they do this to my boy? What was his fault?"

She pushed a rectangular tin filled with papers toward me. The box contained newspaper reports of when Moeen was blinded. In the pictures, his face was badly swollen, the area around his eye blackened. Sabeena's hands were shaking. "Five years ago, if I saw images of violence in the papers, I would turn away. I wouldn't want to see it," she said, obviously holding back tears. "Today, I look. I hope and pray an

Indian soldier has been killed. Maybe it's the same soldier who took my son's eye. I look at the images, and I celebrate. I cook a special meal on those days." As she spoke, her voice quivered with fury and anguish.

By this time, other family members had come into the room too. There was Sadaf, Moeen's younger sister. And there was Mushtaq Sofi, the baker. We all sat on the floor, backs to the walls, filling the small room. The ceiling was barely seven feet high, yet a small ceiling fan spun dangerously in the middle. The walls were painted pink. There were nails hammered into the walls, from which hung large white kurtas, which I presumed were Sofi's. There was a small TV in a corner of the room.

I asked Moeen what he wanted to be when he grew up.

"Cricketer," he replied, shyly. "Batsman."

I smiled, taking notes as I listened.

His father jumped in. "Ask him which team. Go on, ask him."

"Which team will you play for?" I asked, half-knowing the answer.

"Pakistan!" came the response. Moeen and Sadaf giggled.

And why not India?

The two children rolled their eyes. "India is not for us. We are not a part of India, and India is not a part of us. We want freedom from India—*azaadi*—or death. That's all. India's enemy is Pakistan, so Pakistan is our friend."

I somehow knew they would say this, but it came as a shock nonetheless. To hear young children speak of death as if it were something to be welcomed is truly chilling. Their expressions seemed devoid of emotion, their answers almost robotic and rehearsed. But there was an unimaginable pain recessed within, shaping their worldview and thoughts. They had suffered and sacrificed for a fight they barely understood; their sense of loss was intensified by the tales of sacrifice they had heard from friends, family, social media.

I took my leave of the Sofi family. They had been gracious hosts, sharing with me their best biscuits and refreshments. And while they openly told me that Hindus and Muslims could never live together, they had unquestioningly opened their home to a reporter whose name sounded Hindu and who had traveled from New Delhi.

Later, Mukhtar, Arshad, and I wandered back toward where the guards had blocked off the road. Along the way, Arshad stopped by a group of young Kashmiri men. They were dressed in white kurtas. A few had smartphones in their hands. I joined Arshad, hoping to chat casually about their work and their lives. On learning I was a journalist, one of them, a young man named Irfan, began first to talk at me and then to shout. "You people never tell the real stories here!" Irfan's smooth teenaged face had the beginnings of a beard. A small crowd gathered around us. On either side of the narrow alley, old women flung open their windows to listen in. "You want to know why we're angry? I'll tell you," said Irfan, getting more heated. And so it began: independence in 1947, the fraught decision by Kashmiris to join India instead of Pakistan, the promised referendum, the wars, the state of conflict, the impunity of the army, the inheritance of loss, the sense of statelessness, a shortage of jobs, the permanent anxiety and anger. Passersby would stop and chime in. "You call this a life? We're slaves to India. Your whole country is thriving, and we're here living in a hellhole. And then you take away the internet? How dare you?"

There was a certain irony to the fact that we were assembled in a small, curfewed area named Little Pakistan. Apart from me, none of the people there had ever been to Pakistan. I sensed it didn't matter who they supported or what they stood for. All that mattered was what they *did not* feel a part of: India. The frequent internet bans may have prevented getting news in and out, but it had made Kashmiris angrier. As I thanked my interviewees and walked away, I looked back and took in the area one last time. It felt like a tinderbox, just waiting for a spark that would blow the whole place up once again.

Srinagar feels melancholy, but it is stunning nonetheless. The snow-capped mountains are at once forbidding and majestic. The public gardens are a lush, vibrant green, interspersed with spectacular bursts of red and yellow flowers. Dal Lake, near the city's heart, is overrun with weeds and moss but still manages to conjure up visions of a more elegant past. Dazzled by Srinagar's idyllic landscape, one almost forgets how one got there: through a curfewed city, the troubled parts cordoned off, the houses of important people fortified by high boundary walls, iron gates, and a robust military presence.

That evening, I met Zeyan Shafiq, who had become something of a tech celebrity in Kashmir. We had agreed to meet at the café in the Taj Vivanta, a luxury hotel sitting atop a hill with the best views of Srinagar. Shafiq, who was sixteen, was late. He had tried to motorbike to our rendezvous point but was stopped by the police, who discovered Shafiq was too young to have a driver's license. He had to call his father. When Shafiq finally showed up, he was sitting in the backseat of his parents' car, sullen and sheepish, a tech star reduced to an errant teenager.

Mr. and Mrs. Shafiq made me promise to call them when it was time to pick their son up. I agreed.

I sat down with Shafiq in the hotel's lobby, and we each ordered a cup of kahwa, a tealike concoction of hot water, saffron, honey, and almond slivers. Shafiq looked younger than sixteen. He had a soft fuzz on his upper lip. His eyes were a brownish green—unusual in other parts of India but not uncommon among Kashmiris. He stood a little over five feet tall, wearing blue herringbone jeans and an oversized plaid cotton shirt. He seemed to revel in the media attention he had been getting of late. A few weeks before my visit, he had been interviewed by a local newspaper, as well as a radio program. I was his first interviewer from Delhi. He was expecting another from Australia. Perhaps one from London. In April, a few months before we met—when Facebook, WhatsApp, Twitter, and so on were banned in Kashmir—Shafiq had created his own social media platform. It was called Kashbook.

"I don't see what I did as fighting the government," he told me. "I call it helping the people."

Shafiq had first created a version of Kashbook three years ago, when he was thirteen, using basic HTML and Java. It was a fun project but nothing more, and he went on to other things. Then, in 2017, when the ban on social media apps began, he resuscitated it. "Kashmiris weren't able to contact their families. It's a basic need. I didn't want Kashmir to be disconnected from the world."

Within a week of launching in April, Kashbook had signed up fifteen thousand users. Word spread. Every other social media site was banned; VPNs didn't always work. Kashbook was still below the government's radar, so officials hadn't thought to stop it. Suddenly, Kashmiris could once again chat with each other, share photographs and videos. Kashbook became a symbol of the resistance.

"I'm not even against the government, you know," he said. "I'm a techie: I want the support of the government. But yes, once I realized how my site had become a place that was protecting freedom of speech, I knew I had to take it seriously. The internet has transformed India, but my Kashmir has been left behind."

I was struck by the clarity of Shafiq's thought. I had never met a teenager like him before. He described the anguish he felt when the web was completely shut down in 2016. "I can live without food but not without the internet. It felt . . . like . . . ancient times," he said, struggling to come up with a description of his life. "My friends and I would just get bored, sit at home, sleep. It was the hardest feeling in the world."

Shafiq seemed different from the young men I had met a few hours earlier. He was almost apolitical when it came to talking about India or Pakistan; his religion, it seemed, was technology. I asked him about Burhan Wani.

"Kashmiris say he was a hero. I can't call him a terrorist. As far as I've heard, he was a nice person, a sweet guy," he said. "But when he died, everything changed for us. The government began to see social media as something evil, as a tool for terrorism, and for dangerous images to be shared. For me, that was my awakening. It's a red line. You can't ban the internet in a democracy."

India's government had a different take on the matter. In response to growing criticism about the state's internet shutdowns, New Delhi responded by codifying exactly *how* it could suspend telecom services. In a circular dated August 7, 2017, the Ministry of Communications drew on the Indian Telegraph Act of 1885, outlining which state and federal officials had authority to issue a ban.[3] The new laws barely made a splash in the Indian media. There was no public consultation or transparency in the process. New Delhi had made up its mind that for Kashmiris, at least, the web was a privilege, not a right. Few others seemed to care.

Is Facebook a website on the internet or is Facebook *the* internet?

Sheryl Sandberg, the company's chief operating officer, has indicated that both of those statements could be true. "People will walk into phone stores and say, 'I want Facebook,'" she said at the World

Economic Forum's 2015 meeting in Davos. "[They] actually confuse Facebook and the internet in some places."

Sandberg was right. Picking up on her comments—as well as anecdotal evidence from researchers in Asia and Africa—the news website Quartz commissioned a poll.[4] The results were remarkable. When asked if they used the internet, 11 percent of Indonesian Facebook users confidently replied "no"—despite, of course, accessing Facebook through the internet. Nine percent of Nigerian Facebook users came back with the same answer. It was a paradox. A significant number of people in two of the world's seven most populous countries (a) didn't properly understand what Facebook and the internet were and (b) clearly had more brand awareness of Facebook than of the internet.

For Sandberg and Facebook, this should have been welcome news. It is every company's dream to have its product dominate an industry, just as Xerox once became synonymous with photocopying, Hoover with vacuum-cleaning, and Kleenex with tissues. But there was a downside to conflating Facebook and the internet—and it would turn out to have embarrassing repercussions in India.

Facebook had long been approaching saturation. When the social media site hit 2 billion users in June 2017, it was already serving more than half the world's total estimated internet users (around 3.8 billion). If Facebook was to keep growing, the most obvious strategy was to chase the people who were not yet online: an untapped global market of more than 3.2 billion people.

Facebook had already been experimenting with giving people free internet to access its site. In the Philippines, for example, it struck a deal with the mobile provider Globe;[5] the plan was to give mobile users who hadn't paid for data a free pass to connect to Facebook. The logic went that Facebook would offer a gateway to the wider world of the internet. It worked—at least as a marketing ploy. Globe leapfrogged its rival to become the market's number one player, in large part thanks to the promise of free data to use Facebook.

A far bigger plan was afoot, however, called Internet.org. In February 2014, at Barcelona's annual mobile world congress, Facebook CEO Mark Zuckerberg pitched his plan to give people around the world what they really wanted: not just free Facebook but also free internet access. Facebook would link up with phone operators in different countries to provide free data services to users. The goal was to show people what the World Wide Web was capable of, get them hooked, and expand the size of the global internet pie. Everyone would benefit. The biggest catchment areas would be the countries that had the greatest number of citizens still offline, such as Nigeria, Indonesia, Pakistan, Bangladesh, and Brazil. India, as always, was the biggest fish in the pond, with an untapped potential of a billion users.

Zuckerberg would visit India later that year. In October—when the weather tends to be best in the northern part of the country—he made a stop at Chandauli, a village in Rajasthan, descending on the surprised residents in an orange helicopter. Zuckerberg's mission was to understand how rural Indians were using the internet—power outages, primitive handsets, excruciatingly slow speeds, and all. Among the people accompanying him that day was Osama Manzar, the founder of the Digital Empowerment Foundation. Manzar was to later describe what he saw in Chandauli in a column for the Indian newspaper *Mint*:

> What I could gather is Zuckerberg is shy, does not speak much and is still a child at heart, but what was amazing was that whatever little that he uttered, it was about: how to bring service-oriented information to people through the mobile as a device; how to bring high-speed Internet to the people; how to offer services and content as per daily needs through apps; and how to enable people living in remote areas so they can share their culture, knowledge, products and services with the rest of the world.[6]

Zuckerberg also stopped in New Delhi, where he met Prime Minister Narendra Modi and keynoted an Internet.org summit at the Taj Mahal Hotel. At his speech—which I attended, along with several local reporters—Zuckerberg explained why he wanted to help proliferate internet access across India: "It should be like dialing 911." A public necessity, in other words. (It was beside the point that India does not have an equivalent for dialing 911; Indians got the reference.)

Facebook had recently been citing data to back its vision of the internet—and Facebook—as a public service. A 2015 Deloitte study showed that the social media company had "enabled $227 billion of economic impact and 4.5 million jobs globally in 2014."[7] Deloitte also claimed that if India reached Western levels of internet access, it would be able to create 65 million new jobs and slash extreme poverty by nearly a third. (Facebook, which of course embraced the study, often neglected in its outreach to highlight one key point here: it had commissioned—and therefore paid for—Deloitte's research.)

Zuckerberg was getting front-page treatment in India. Here was a bona fide tech superstar doing the rounds and burnishing India's credentials as a new tech frontier. The Modi government, then newly in power, was only too eager to open doors. Modi himself had more than 30 million "likes" on Facebook at the time (and has since crossed 43 million—four times as many as Tom Cruise, and twice as many as Bollywood's Shah Rukh Khan).[8] India's government was also planning its own Digital India campaign, which aimed to put essential government services online. All the signs pointed to Internet.org gaining significant traction in India. What was not to like? India needed to get people online—and here was the world's number one tech titan putting its weight behind the effort. Facebook was tying up locally with Reliance's Jio, which was readying to roll out its massive free 4G mobile plan across the country.

There was a catch. The free internet on Internet.org was not really the full World Wide Web. It was Facebook's web, a sort of internet-lite, with access only to about three dozen websites: one search engine (Microsoft's Bing, not Google), a few general-interest sites, and—of course—Facebook.

At a press conference after Zuckerberg's Delhi keynote, reporters were lining up to ask him about Internet.org. Even among a generally adoring press corps, there were the first murmurs of skepticism: Would Facebook's gated internet benefit a few preferred partners? Would it violate net neutrality—the principle that internet providers should treat all data on the web equally? What were the longer-term implications of having Facebook dictate what kind of internet people discovered? Those questions never got asked. Facebook's local PR people had preselected questions they wanted asked, such as those involving the positive effects of access and the speed at which this would happen. Net neutrality was simply not on the table.

A few reporters were irked by this attempt at control. One young journalist in particular made it his mission to make his questions public:

Nikhil Pahwa, the founder of the technology website MediaNama. Pahwa is a self-confessed workaholic. An engineering school dropout, he found a job at a website called ContentSutra, fell in love with the idea of writing about technology, and then decided to strike out on his own. "I wanted to do the most difficult thing I could do," he told me. There were few more challenging assignments than running a media company. Founded in 2004, more than a decade later, MediaNama had become a profitable, sustainable business, featuring smart tech journalism and hosting regular events across India (most of the group's income comes from its sponsored events). Pahwa is the group's lead writer, editor in chief, ad salesman, and spokesman rolled into one.

"Not a single international company's CEO has ever come to India and agreed to speak with me," lamented Pahwa. "Are they scared? I guess it's because I'm a difficult guy who asks difficult questions."

The morning after Zuckerberg's Delhi 2014 keynote, Pahwa took to his website to vent his thoughts, writing that "what Zuckerberg means by 'internet for all' is essentially Facebook for all, along with a few non-profit services thrown in to give it the appearance of philanthropy."[9]

Pahwa worried that Facebook would have too much power. What would happen if Internet.org offered access to WhatsApp (owned by Facebook) but not Hike, a rival messaging app? What would happen if Alibaba was on the gated Internet.org but not IndiaMART, Amazon, or Flipkart? "Internet.org makes business sense for Facebook, but not for India," concluded Pahwa.

Others were beginning to ask the same questions. Even Osama Manzar, whose life's work was to bring the web to the farthest corners of India, was souring on Facebook's plans. "The whole thing was about launching their product—not the real internet," he later told me. Manzar had been interviewed several times by Facebook executives in the months leading up to Zuckerberg's India visit. They had been picking his brain. He hadn't realized the company's offering would be so limited and self-interested.

There was a larger issue at play here. Zuckerberg's Internet.org announcement came at a moment when internet activists in India were arguing with a cartel of telecom operators about the very topic of net neutrality. In February 2015, Facebook went ahead with its launch of Internet.org in India. A month later, the Telecom Regulatory Authority of India (TRAI) announced it would soon make a ruling on net neutrality—a ruling that would now also affect Facebook.

By now, Pahwa was working full-time on a campaign to ensure net neutrality was protected in India. It had become his sole focus. He had delegated the daily running of MediaNama to his team. "As soon as I heard that TRAI was going to rule on this, I mailed the guys at AIB," he said, referring to All India Bakchod, a group of comedians and satirists known for their social media savvy. He wanted them to produce a segment on net neutrality along the lines of a takedown by John Oliver of HBO's *Last Week Tonight* fame. "I told them: this is going to kill the internet in India. We have to fight for it."

Fight they did. AIB's simple—and hilariously funny—video explainer gained instant viral fame when it was posted on April 11, 2015.[10] It got 3.8 million views and directed people to voice their support on SaveTheInternet.in, a website built by Pahwa and a group of activists. Suddenly, an arcane topic like net neutrality had hit the mainstream of debate on TV and social media.

Meanwhile, Facebook was planning a response. As the investigative journalist Rahul Bhatia has reported in a comprehensive long essay for the *Guardian,* Facebook executives had already launched a significant lobbying campaign in the corridors of power in Delhi.[11] Internet.org was rebranded as the more humble-sounding Free Basics. When TRAI announced it would reach a decision on net neutrality in December 2015, Facebook stepped on the gas. Users who logged on to Facebook from India saw a specially planted corporate message, warning that TRAI could ban its plans "to [connect] one billion Indians to the opportunities online" and inviting people to click on the message and send an automatic email to TRAI in protest. Facebook was simultaneously bombarding newspapers and billboards with advertisements. "This was *our* fight—India's fight—for net neutrality," recounted Pahwa. "It was our debate. These guys [Facebook] came in and made it all about *them*."

If it was starting to feel like Goliath versus David—the founder/CEO of a company worth $500 billion versus the founder/CEO of a company worth less than 0.001 percent that amount. That feeling was cemented when Zuckerberg and Pahwa dueled in opposing editorials in the *Times of India* two days before TRAI's deadline for public comment on net neutrality. "Who could possibly be against this?" wrote Zuckerberg, boasting that 15 million people had already come online through Free Basics.[12] Nearly half of those users had paid to access the full internet within thirty days, he claimed. "Choose facts over false claims. Everyone deserves access to the internet. Free basic internet services can help achieve this. Free Basics should stay to help achieve digital equality for India."

Pahwa wrote a scathing rebuttal, pointing to other programs: Aircel was offering full internet access for free at low speeds; Grameenphone in Bangladesh was giving people free internet if they watched an advertisement.[13] Free Basics, said Pahwa, would give Facebook the authority to reject websites and apps from joining the program. "Facebook is being disingenuous when it says that Free Basics is in conformity with Net Neutrality," he wrote. "[This] can be harmful for India's democracy. It is a form of vertical integration that is anti-competitive and inimical for India's fledgling start-up ecosystem."

The closing arguments had been made. TRAI—judge and jury—deliberated for months. When the decision arrived in February 2016, it was definitive: net neutrality needed to be preserved in India. Put another way, Free Basics would have to go. Facebook had lost.

"It felt like a small band of us nothings, up against the most powerful group of people in the world," Pahwa told me later. "And we won. We protected net neutrality in India."

Sometimes the little guy does win. For all India's many ills, the country's regulators and courts have largely protected the rights of the average citizen. One example of this came in February 2017, a year after the net neutrality ruling, when India's Supreme Court began deliberating a question so typical of the digital era: Was privacy a fundamental right for 1.32 billion people?

It was David versus Goliath redux.[14] On one side, a small group of petitioners argued that the biometric data collected for the government's Aadhaar identification program violated an individual's right to privacy. Defending itself against that charge was the government itself, arguing that the constitution did not even *recognize* privacy as a fundamental right.

On August 24, 2017, the Supreme Court announced its ruling. Drawing on the writings of John Stuart Mill and James Madison, the judges made a case for enshrining privacy as a fundamental legal right in the modern digital era.

"This is not just a legal victory. It is a moral victory," Nikhil Pahwa told me in a CNN interview. Pahwa was not campaigning directly this time, but he was a vocal supporter of the right to privacy. India, in his view, now had a more progressive set of laws on privacy and net neutrality than many of the West's most developed economies.

Within minutes of the Supreme Court judgment, the opposition party's leader, Rahul Gandhi, tweeted that the ruling was "a major blow to fascist forces"—a reference to the government's attempts to make signing up for Aadhaar mandatory.

The government was quick to punch back. At a hastily organized press conference an hour after the ruling, Law and Justice Minister Ravi Shankar Prasad said Gandhi "needed to do his homework," suggesting his younger adversary hadn't read the whole ruling. (This was likely: it ran more than five hundred pages.) Prasad went on to read a summation from the ruling. "The right to privacy is not absolute. It is to be determined on a case-by-case basis," he said, peering down at a stack of papers in front of him, as cameras carried his message to millions of TVs across India. "The whole world is marveling at our technology," he continued, referring to the fingerprint and retinal scans used in the Aadhaar database. "We take minimum information for maximum use."

Activists disagreed. "Data is the new oil," said Pahwa. "There has been a move to treat citizens and their data as a commodity that can be sold. But once you've given up your data you have no rights to it. What this judgment has done is give power back to the citizens of India."

Armed with the Supreme Court's ruling, activists were now hoping to take on the crux of the issue itself: Aadhaar and whether signing up should be mandatory. "I don't have an Aadhaar number and I don't want one," announced Pahwa. "I don't want to be under surveillance and have my every move be documented."

The government's comeback is that the goal isn't surveillance. Instead, the Modi administration has coined a pithy acronym called "JAM"—Jan Dhan, or bank accounts; Aadhaar, or identification; and Mobile—to symbolize how it plans to link three vital services. JAM would help in creating a social welfare policy to enable the state to directly transfer funds to those most in need.[15] It would also ensure that subsidies would reach citizens directly, and with no chance of middlemen taking cuts. To make all this happen, the argument goes, one would need to sign up for Aadhaar.

The ruling didn't challenge the constitutionality of Aadhaar itself, but it did open a door to challenging the unfettered collection of data. A "fundamental right"—which the court had declared privacy to be, despite what Prasad intended—technically superseded any other rights given to Indians by law. Now, in theory at least, citizens could take the state to court on the grounds of privacy and question the collection of data. But they would be in a minority. More than a billion Indians have already signed up for Aadhaar, handing over their biometric data to the state.

The stakes are high. There have been several reports of personal data leaking from the Aadhaar database. In January 2018, *Tribune* journalist Rachna Khaira reported that she had found people who were selling access to the Aadhaar database.[16] For a small fee of 500 rupees, one could get access to an Aadhaar portal; once in, you could enter anyone's Aadhaar number and be able to access the person's complete details—name, address, phone number, and email. For a further 300-rupee fee, one could print out a particular person's Aadhaar card, which in turn would provide access to other forms of identification and accounts. But in a twist, once Khaira's story was published, the Unique Identification Authority of India—the group that runs Aadhaar—issued a police complaint against not only the sellers of the data but also Khaira. It was a classic case of shooting the messenger. Edward Snowden, former CIA employee turned whistle-blower, tweeted in support of Khaira: "The journalists exposing the #Aadhaar breach deserve an award, not an investigation."[17]

"The government has failed us repeatedly," Pahwa told me. "It's downright lies when the government says Aadhaar is safe. Anyone has access to the data. Lies can only last for so long." And any leaks were permanent: unlike a password, fingerprints or retinal scans could not be changed.

"People in tech just foolishly assume that the government is going to do the right thing," added Pahwa. "But the one thing you know about the Indian government is incompetence. We're only just realizing how much of a mess we have created here in India."

Zeyan Shafiq loves Kashmir dearly, but he knows he has to leave—possibly emigrating to another country—to go to university.

"This is the most beautiful place in the world," he told me. "But if you're ambitious, this is no place to be anymore."

A few months after I first met Shafiq, he had lost control of Kashbook. His cofounder had ousted him from the team over a petty dispute. Shafiq was upset, but he had moved on. Now that the specific ban on social media sites was over, people had gone back to using the real thing—Facebook—as well as other apps like Instagram and WhatsApp. But the government could still shut down the internet at any time.

"My aim is to learn from the best in America," he told me. "Why can't I be a Mark Zuckerberg, too, and create something the whole world knows about?"

I asked him if he thought that America's 2017 decision to reverse net neutrality made India a more progressive market when it came to freedom on the internet. Yet as I asked the question, I winced, instinctively knowing what his answer would be.

"You call this freedom? Facebook might have been trying to bully India. Okay, maybe. But the biggest bully in India is the government itself. And until they don't stop shutting down the internet, we have no stability or security here," he said. "I could never start a self-respecting internet company in Kashmir. How can you explain to your customers that 'we're sorry, today we can't work for you because they shut the internet down'? It's ridiculous. My only hope is to get out and see if I can do something that might help from outside."

8

The Great Indian Currency Scramble

Digital Money

AN HOUR BEFORE DAWN on November 8, 2016—a Tuesday—Sarvesh Kumar woke abruptly from what had been restless sleep. As he propped himself up on his *charpoy*, he told me, his eyes adjusted to the darkness of his tiny room, tucked away in a slum near New Delhi's prosperous Vasant Kunj neighborhood. It was, he recalls, pitch black. Kumar's inability to see enhanced his other senses. His ears picked up a chorus of mournful howls from the stray dogs roaming the streets some distance away. A few seconds later, he began to take account of a gentle rumble of snoring. Bandhana, his wife, was still asleep beside him. The next stimulation to hit him, he said, was that of smell: they had saved leftovers from dinner in a small dish, placed on the floor in one corner of the bedroom. (There was no kitchen, no fridge, no cupboards.) The smell of stale dal and starchy rice lingered in the air, mingling with the stench of urine from the latrine right outside. The apartment's only toilet had no working flush, just a mug and a bucket with which to pour water. At night, Kumar's father—who slept in the next room—would get up to urinate in the dark, sitting on his haunches and peeing all over the cramped toilet floor.

Kumar was only twenty-two years old but looked much older. He was developing a hunch. His gray shirt and trousers—both made of the same rough, fraying cotton—seemed to fall over his limbs like oversized bags. Life was wearing him down. Kumar was working sixteen-hour shifts driving a three-wheeled auto-rickshaw, seven days a week. Ever since he had gotten married and brought Bandhana to join him in the city, his expenses had ballooned. His father, meanwhile, was working less and less, his health and eyesight beginning to fail him. Asha Ram Kumar worked as a chauffeur for a family nearby in Vasant Kunj. He had recently bumped their car into a lamppost. Any day now, Kumar told me, his father would lose his job.

The year 2016 marked Kumar's eighth in New Delhi. He had left the family village near Farrukhabad in Uttar Pradesh at fourteen to join his father in the city. There had been no work back home. The ancestral plot of land had become barren: it had already given too much to too many. Like others before him, Kumar had to leave home to grow up. In the city, he learned to hustle. He bribed a local dealer to issue him a license to drive three-wheelers. At the age of sixteen, Kumar began to work a full-time job driving an auto-rickshaw.

That was the work he continued to do, day in, day out. He was illiterate but was soon schooled in the ways of city life, making more money than he could have imagined possible in the village. On a good day, he would pull in a thousand rupees in fares—about fifteen dollars. But his expenses also soared. Nearly half of his take went to rent his rickshaw and pay for gas; another third went to his landlord to rent the two-room apartment he shared with his father. After costs for food and household supplies, there was little to send back to his mother and sister in the village. He had no savings, no bank account.

Kumar had begun to give up religion after he moved to Delhi. "God doesn't have time for me, so I don't have time for God" was the expression he used. His father, he told me, was the same. The Kumars followed tradition in the village but adopted different personas when they were in the city. There were, however, childhood superstitions even the younger Kumar couldn't shake off. Tuesdays—Mangalvaar in Hindi—were days that inspired in him a strange, almost spectral fear; as a child he had learned from his mother that Mangalvaar was the domain of the god Mangal, or Mars, the troublemaker. His mother would fast on Tuesdays to ward off the evil eye.

With a sense of foreboding, Kumar decided he would try to get another hour's sleep. At dawn, he would have to go and clean his auto-rickshaw before heading off to find his first passengers.

That same Tuesday, halfway around the world, it was Election Day in the United States. In an increasingly polarized electorate, it seemed left and right were casting their ballots against the other as much as they were voting for their own candidate. Trump versus Clinton was a dramatic and bitter showdown—the most expensive election in history—and it had consumed global media.

Even in New Delhi, there was much speculation about which candidate would be better for India. What would it mean for Pakistan? Would the next occupant of the White House support India's quest to join the United Nations Security Council? I found myself caught up in the ritual punditry as well. As a journalist representing CNN in New Delhi, I was often called to participate in panel discussions and TV news debates, especially when the topic had anything to do with American politics. That evening—morning in the United States—was expected to be a blockbuster night for Indian TV news. Channels were planning to broadcast state-by-state results through the night and into the morning. I agreed to join a TV discussion on our Indian affiliate CNN News18 at 8:00 p.m. My copanelists were Meera Shankar, a former ambassador to the United States, and the well-known Indian editors Vir Sanghvi and Pankaj Pachauri.

As we sat down around the anchor's desk, however, we were informed there would be an interruption. TV news channels had just received notice that Prime Minister Modi was to deliver a major announcement. My fellow panelists joked that perhaps Modi was trying to steal the media spotlight away from Washington. We wondered what could possibly be so important.

We were soon to find out. At 8:00 p.m. sharp, the nation's TV broadcasts cut away from US election coverage to show Modi standing alone at a lectern with a microphone. He was speaking in Hindi. "India has become a bright spot in the global economy," he declared in a sonorous tone. "This government is dedicated to the poor. It will remain dedicated to them."

Listening in the studio, we were mildly confused as to why Modi was speaking in vague generalities about the economy.

"In the past decades, the specter of corruption and 'black money' has grown," he continued, referring to incomes that were undisclosed and therefore avoided taxation. "It has weakened the effort to remove poverty."

This was a frequent refrain, and we all wondered how long the TV channel would continue broadcasting the speech, given its seeming lack of newsworthiness.

"There comes a time in the history of a country's development when a need is felt for a strong and decisive step. For years, this country has felt that corruption, black money, and terrorism are festering sores, holding us back in the race toward development."

We began to get restless in the TV studio. Surely it was time to cut away and go back to discussing the biggest election of the year?

Then came the bombshell.

"Brothers and sisters, to break the grip of corruption and black money, we have decided that the 500- and 1,000-rupee currency notes presently in use will no longer be legal tender as of midnight tonight, that is, 8 November 2016. This means that these notes will not be acceptable for transactions from midnight onward."

We looked at each other, stunned. The 500- and 1,000-rupee notes were the two highest-denomination bills in Indian currency. Together, they represented 86 percent of the total value of cash in circulation. And India, unlike Western countries, was mostly a cash-based economy: 98 percent of transactions—by volume—were carried out in cash. Less than 2 percent of the population had credit cards. (India also had a ridiculously low 693 card-reading machines per million people. By comparison, China and Russia had 4,000 point-of-sale terminals for every million citizens.) India simply could not survive without paper cash.

And yet, as quickly became evident, we had indeed heard correctly. Modi was outlining the details of how people could return their old cash to banks, and withdraw newly designed 500- and 2,000-rupee notes. Introducing new bills from time to time is standard in most economies, but not in this way, with the two biggest currency notes impacted, and no prior notice. Instinctively, we checked our pockets to see how much cash we were carrying. Several of our notes would soon be out of use.

As Modi kept speaking, we were gently ushered out of the TV studio. Realizing the gravity of the story, the news channel was frantically booking economists to try to make sense of the announcement.

Modi had prefaced his announcement by calling it a "decisive step." That was an understatement. India was about to be hit by a man-made financial tsunami. The media quickly gave the plan a name:

"demonetization." Suddenly, no one seemed to care who won the US presidential elections. The real story was right at home. Word about the cash ban spread like wildfire. By 9:00 p.m., as hundreds of millions of people rushed to assess their finances, India was jolted into a hive of financial activity. A great nationwide scramble had begun.

As I traveled back home that night, the streets were abuzz with concern and chatter. I stopped at a couple of ATMs but found them shuttered. People had gathered at their local markets and stores to get more information. No one I spoke to seemed to understand the scope or scale of what had just been announced. There seemed no way that 86 percent of India's currency could be returned to banks—and new bills distributed—without mass confusion and pain. The farthest, disconnected parts of the country would undoubtedly suffer. There was also the question of what would happen to the millions of Indians without bank accounts, and without the ability to exchange their old notes for new.

Meanwhile, government talking heads were already seizing the narrative on India's many nightly TV talk shows, calling Modi's decision a masterstroke that would finally end corruption in India. The theory went, said the talking heads, that people who had evaded taxes would be caught with bundles of cash and nowhere to put it. There were also several fake news reports, of the kind I alluded to in Chapter 6. Some TV channels were reporting that the new 2,000-rupee notes would come with GPS-enabled microchips ingrained, to catch people with too much money. The reports turned out to be science fiction.

As Modi himself pointed out, one of the intended targets of demonetization was "black money"—cash that hadn't been declared to the tax authorities. But black money was everywhere; it was self-perpetuating. In India, everyday transactions are conducted in cash without receipts—buying vegetables, visiting the dentist, filling up gas. Customers oblige because the transaction is off the books. The buyer avoids sales tax, and the seller has less income to report. Sometimes businesses put a transaction on the books but find other ways to deceive the taxman. Saree shops, for example, often put an implausibly low sticker price on their wares and in their books. When it's time to sell, they add a zero to the price, showing it to the customer on a calculator. The shopkeeper will then hand out a receipt that is worth 10 percent of the actual cash transaction (and therefore impose a sales tax that is a tenth of what it should be). Shopkeeper and customer both avoid their respective taxes. Home sales follow a similar model. Apartments in the big cities usually include a "white money" sticker price, with a far larger black money component to be paid purely in cash. The system works because an alarming number of people are complicit. Everyone manages to skirt the full sales or capital gains taxes—and because the transactions are largely in cash, they can also avoid disclosing an income or expense on their tax returns. The system perpetuates more of the same.

While tax avoidance is common at every level, the real beneficiaries are India's rich. About 95 percent of India's population earns less than the minimum 250,000 rupees annually (about $4,000) to qualify for the lowest income tax bracket. Put another way, for more than a billion Indians, black money is just a fantasy. They do not make enough to accumulate illegal wealth. But the remaining 5 percent—India's income tax payers—were the stated targets of Modi's demonetization. Most of all, businessmen with large stashes of cash—and without receipts or income tax statements to paper-trail the money's provenance—were in trouble.

Suddenly, a race was under way to launder this black money. One jeweler in Delhi later told me that he invited his most loyal customers to his home so they could exchange cash for jewels before the midnight deadline. (Unsurprisingly, the jeweler refused to be quoted. He had sold expensive baubles at a premium and later needed to find ways to launder his own cash.) An executive for the international luxury group LVMH recounted how its Louis Vuitton stores were inundated with unusual requests that night. People wanted to buy store credit worth hundreds of thousands of dollars. (LVMH did not accept these deals.) India's rich were everywhere on the night of November 8, begging dealers to sell them cars, homes, anything. Reports would later emerge of households paying their drivers, maids, and guards salaries up to a year in advance, just to get rid of their cash hordes. Since most Indian workers get paid in cash, these transactions were also off the books. Ninety percent of India's workers are part of the so-called informal economy—a euphemism if there ever was one—where there are no labor contracts, job security, pensions, health care... only cash payments.

As dawn broke on November 9, I scrambled to cover the story for CNN. The best visual backdrop to our live programming was obvious: banks and ATMs. I had never seen anything like it. At New Delhi's Connaught Place, a giant colonial-era market with buildings that clustered in concentric circles, one ATM attracted a queue that snaked hundreds of meters long. It was bedlam. No one seemed to know what was going on. A few people had managed to get the new 2,000-rupee notes. However, the machines weren't equipped to dispense the new 500-rupee notes, which were smaller in size. Moreover, the stock of 2,000-rupee notes was not large enough to satisfy demand. By 10:00 a.m., the ATM had run out of cash. Tellers at the bank adjoining it were ill equipped to deal with the hundreds of people who had queued up. And this was in central Delhi. Reports emerged of similar scenes across India's cities, broadcast live on the country's news channels. (It would take longer for the media to get a better sense of the impact in rural India, where a cash recall would prove to be more debilitating.)

I spoke with dozens of men and women lining up to withdraw the newly issued currency. Some had been waiting for several hours. Yet most seemed content.

"Modi's done a great thing," said Abhishek Tewari, a twenty-six-year-old call center worker. He had worked all night and was bleary eyed, but the adrenaline was keeping him going. "This is a pain for us, yes. But we are honest, salaried people. We won't lose anything. The rich business families with black money are going to suffer."

Others around him chimed in. "This is a huge inconvenience. But we'll manage. This is good for India," said a schoolteacher who worked nearby. "Finally, something is being done to fight corruption," she added.

The long lines at banks and ATMs would continue for several weeks, as the government extended the deadline for returning old notes to December 31. Up to 250,000 in old notes could be accepted per person; for any amount over this, people would have to explain the source of their money.

New notes remained difficult to come by. In cities and towns across the country, millions waited hours for their turn to take out a maximum of 20,000 rupees a day, the limit imposed to prevent citizens from stockpiling new notes. Meanwhile, some media reports told of people who died while waiting in the bank lines. (I wasn't comfortable with running those stories, which were intended to dramatize the considerable—and unnecessary—pain the government's plan was causing.)

Wealthier citizens, however, didn't seem to be languishing in these bank queues. A Delhi-based start-up called BookMyChotu had become instantly popular. The company's website, BookMyChotu.com, encouraged people to reserve one of their "Chotus" to wait in line for them at an ATM. (*Chotu*—literally "small guy"—is a derogatory Hindi term for a manservant.) The rates ranged from 90 rupees an hour to 500 for seven hours in line: small change for Delhi's upper classes. In a sense, BookMyChotu only formalized what had already been happening. India's wealthy always have a man Friday to run their errands.

Over the next few days, news reports focused on how India was coping. Weddings, which depend heavily on cash payments for vendors, were disrupted. There were reports of people being turned away from hospitals without evidence that they could pay their bills in cash. Businesses seemed to be running on a barter system, goods for goods rather than goods for cash. Daily wage workers were going unpaid. India's financial arteries were being sapped of their blood. Chaos reigned.

For several establishments, there were severe financial consequences. Business at a mom-and-pop store in Delhi's Vasant Vihar market was down by 90 percent. "No one has cash," the owner told me. He had been running his small convenience store for thirty-five years, selling essentials like milk, bread, and toothpaste. "I never installed a card machine, and now I'm unable to sell anything to people," he said mournfully. "How will I feed my family?"

At an auto-rickshaw stand near Vasant Kunj, I spoke with several drivers who said their incomes had been halved. It was here that I first met Sarvesh Kumar.[1] He was shy; when he stood up he revealed himself to be a little more than five feet tall. But even with his large, sorrowful eyes he had a determined look on his face. If it weren't for his slight stoop and his unshaven face, one could easily mistake him for a boy.

"People are getting into my auto for rides, but when I take them somewhere I always learn they don't have enough cash," he said to me in Hindi, with an accent so pure that I imagine he used it growing up in Uttar Pradesh. "I thought demonetization wouldn't affect me, because I never get to see the big 500 and 1,000 notes anyway. But this has put a squeeze on the smaller bills as well. No one has change."

"Have you tried Uber?" I asked. The ride-sharing apps Uber and Ola had seen a surge in popularity in the last week, as more people booked cars on their platforms to avoid having to pay cash for a regular taxi.

Kumar had heard of app-based taxis. Some of his former rickshaw-driver colleagues had signed up. He lamented it wasn't for him: his phone was an old-school feature phone without apps. And crucially, he was completely illiterate. Kumar couldn't even read Hindi, which would have enabled him to use Ola.

Demonetization had been in effect for nearly a week when I first spoke with Kumar. He could no longer afford to bring his wife some fruit in the evenings. They were eating just rice and dal at home—no vegetables to go with it. There was no way he was going to make rent.

"What will you do?" I asked.

"I don't know," he responded, as we chugged around Delhi in his auto-rickshaw. "Maybe they'll let me off this one time?"

I didn't share Kumar's optimism. Yet what struck me the most was his complete and utter faith that demonetization would turn out to be a good thing for India.

"This is *really* going to hurt the fat cats!" he exclaimed. "See, I know how to deal with hunger for a little while. My income will return eventually. But what about the rich? They don't know how to sustain pain like we do. Now they'll suffer!" He chuckled.

I told him about BookMyChotu and reports I had heard of India's rich finding ways to launder their untaxed cash. Which of those "fat cats" would go hungry like he did? Which of them would struggle to make rent?

Kumar nodded thoughtfully, but his mind was made up. He told me of reports he had heard of a rich businessman in his neighborhood: a tip-off had drawn the police to his house, and they found millions of untaxed rupees and jewels hidden in a locker. Now he was in jail.

The question was how many of these kinds of people would get caught. Kumar didn't seem to care. I realized that for him, the mere thought of chipping away at the edifice of India's rich gave him such joy that he was willing to countenance hardship to his own family. I suspect he knew it was an irrational feeling. It didn't matter. It was like a powerful, life-restoring drug. This was payback for everything he had ever suffered.

Modi had seemed to put his finger on the poor's greatest yearning: ending corruption. It was as if this one issue would unlock solutions to all of India's problems: job creation, inequality, bad schools, a shortage of doctors and judges. Whether demonetization would work or not, people like Kumar seemed willing to take a collective leap of faith.

I wondered at the time how long Kumar's optimism could last. How long would it be before his schadenfreude turned sour, when his own suffering exceeded the pleasure he got from watching the rich squirm a little bit? Most of all, I wondered how many other Kumars there were across India, particularly those far from the big cities who had no access to the new rupee notes, to banks, to a daily wage. Once India's cash-recall tsunami had subsided, what kind of devastation would be left behind?

When Modi rolled out his demonetization plan, his stated focus was to hit corruption; he also said he wanted to stymie groups that were funding terrorists—with either real or counterfeit currency notes. But as the days wore on after November 2016, government spokespeople began to adopt a different narrative. The real motivation behind demonetization, the grand master plan all along, was to push India toward becoming a cashless economy. The new mantra was *digitization*, part of the government's Digital India campaign. Modi's shock cash recall had turned out to be a massive boost for India's fledgling digital wallet companies, and the government wanted to take credit, as it were.

The biggest winner was a start-up called Paytm, or Pay through mobile. "We used to sign up a few thousand customers a day, but now we are up to half a million daily," Vijay Shekhar Sharma, the founder and CEO of Paytm told me at the end of November 2016.[2] I was interviewing Sharma at the group's headquarters in Noida, a suburb just outside Delhi. In the three weeks since November 8, Paytm had gained 10 million new users—an annualized 54 percent growth rate. The speedy adoption of digital wallets was made possible entirely because of the mass proliferation of internet-enabled smartphones. Paytm's mobile app had a simple interface. One could sign up with a mobile phone and transfer money into a personal digital wallet through a bank account. The wallet could then be used at stores (or online) by scanning a QR code or punching in the recipient's phone number. Transactions were quick and easy, even with a limited understanding of technology. It was similar to PayPal and Venmo in the West, with the difference that almost all the users had signed up on mobiles and not PCs.

Sharma was part of a new breed of Indian tech icons. Unlike business leaders from a generation earlier, he didn't grow up in Anglicized privilege. He didn't attend an elite urban school or inherit a large family business. Instead, Sharma grew up in a small town near Aligarh in Uttar Pradesh. He went to a Hindi-medium school, with subjects like history, geography, and the sciences taught in Hindi—not English. Sharma's father taught at a different school, drawing a small salary that was just enough for the family to lead a middle-class life. Despite his modest beginnings, Sharma was ambitious. He graduated at the top of his high school class and dreamed of studying at the elite Indian Institutes of Technology. His hope was that he would then go on to the United States and attend Stanford University, given the number of tech millionaires that it had produced. When Sharma looked at the requirements to enter the IITs, however, he realized that he needed to be proficient in English. There was no point even attempting the entrance examinations. Instead, Sharma traveled to the capital and found himself a place at the less prestigious Delhi College of Engineering. "I could barely understand what they were teaching in class," recounted Sharma, laughing. The physics and mathematical terms in English made no sense to him. But Sharma was determined to make it. He was known to read a book in Hindi, and then read the same book in English in an attempt to teach himself the language.

It remained a tough journey upward. Sharma learned to code and built a content management system. He signed up important clients. Even then, Sharma struggled to make ends meet, flirting with bankruptcy and ruin. His salvation turned out to be an idea whose time he believed had come: a bank that would be accessible for average Indians on their smartphones. Perhaps unsurprisingly, his investors believed him when he said India was ready for this new venture. The idea was Paytm.

"Why are we successful? Trust." As he was speaking to me—on camera, and in the middle of his open office floor—a group of workers in the background burst out laughing at some joke. The surge in sound was picked up by our microphones. Sharma leaped up and waved his arms in the air. "*Hello!*" he shouted. "Interview going on. Please no noise, please no noise," he called out, adopting what seemed to me the singsong voice of a school principal speaking to first graders. There was obvious camaraderie between Sharma and his team. In the evenings at the office, he would play music on the company speakers, turning up the volume as the evening progressed. (Sharma was Paytm's resident DJ. Coldplay and U2—his favorites—were on the playlist when I was there in November.) Sharma sat back down to continue the interview, not skipping a beat. "Many others have a similar service to ours, but we have a special relationship with our customers. We understand who they are. We listen to them."

Trust has long been a formidable hurdle for banking services to get over. Indians often worry about fraud when it comes to card-based payments. Seventy percent of online shopping is paid for in cash at the time of delivery; about that many app-based taxi payments are also made in cash. Putting too much money into one card account seems dangerous to the average Indian—what if the state cracks down on the rich? India remains a place where trust in large systems is limited. Instead, Indians tend to trust family and like-minded friends they can hold accountable, and while that will change over time, the change will be gradual.

I wondered whether some of Paytm's secret was the simple fact that Sharma was an everyman like his customers. Even at forty, with a 20 percent stake in a start-up that was now valued at several billion dollars, Sharma dressed casually in an unremarkable blue shirt and black trousers. There was nothing fancy or affected in his demeanor. He spoke in crisp English—which he had taught himself—but frequently broke into Hindi with me and with his colleagues.

"We know the needs of the people. We *know* our people," Sharma said, responding to my question about his mass Indian appeal. "Four out of seven of our customers are rural. That's because in rural India you don't have so many places where you can use a debit or credit card."

Sharma had noted that India had a shortage of card terminals at stores across the country. By his count, there were only 1.5 million such point-of-sale devices in the country, often concentrated in shopping malls.

"We have signed up 1.5 million new vendors in ten months," he said. In other words, Sharma had accomplished in under a year what the entire credit and debit card industry had taken decades to do. His strategy was simple: hire an army of freelance salesmen who could explain to small business owners why Paytm would be useful for their businesses. "Even now, we are signing up about seventy thousand new merchants a day. Demonetization has brought a real attention to the fact that mobile is the only way forward in India."

I had seen much evidence of Paytm's drive in the days after November 8. My local dry cleaner—who had no card machines at his store—had recently signed up. Gas stations, restaurants, even roadside *dhabas* had begun to turn to digital wallets. The Paytm stickers on storefronts had become ubiquitous.

"Look at how easy it is to use WhatsApp," Sharma pointed out. "That's why they have some 200 million users here. We put in a lot of effort to train people how to use Paytm. This is India's occasion to rise up and become a digital-native country." (About a year after Sharma and I spoke, WhatsApp added a banking feature to its app, thereby becoming a direct competitor.)

The numbers bear this out. Paytm's user base grew from 122 million in January 2016 to 218 million by March 2017. The app recorded 1.5 billion transactions in the 2016–17 fiscal year, worth more than $5 billion. Other digital wallet companies were booming as well. MobiKwik—founded by an IIT graduate, of course—had seen its user base grow 70 percent to 55 million in the six months following demonetization. For each of these transactions, the company would charge only a tiny fee, but given the volume the numbers quickly mounted.

Curiously, one mobile banking service declined in users after Modi's cash recall—at least temporarily: MoneyOnMobile, which a year earlier claimed to be India's largest mobile payment platform. MoneyOnMobile has a different model than digital wallet companies. While Paytm was targeting users with smartphones, MoneyOnMobile's typical user lacks a smartphone or a bank account. In the money transfer industry, MoneyOnMobile's business is an example of the "assisted model": users typically show up at a small store that runs the MoneyOnMobile platform and ask the store owner to process a transaction for them. The owner's "assistance" is necessary in part because the user is either illiterate or unable to use technology.

In July 2017, I traveled to Kolkata in West Bengal to see how it works. Along the Dock Eastern Boundary Road in the city's south, right off a street named after Karl Marx, I found the unremarkable-looking Shaw Auto Parts store. An electricity pole seemed to grow out of the pavement in front of the store, with wires emerging haphazardly in several directions. The pavement was narrow, broken into small blocks of concrete. Inside, Shaw Auto Parts was dingy, lit by a single bare lightbulb, and quite small, about five feet wide and six feet deep. Lining its walls at the back was an array of used metal parts, greased and coaxed back into service. The shop was named after Guru Prasad Shaw, who had migrated from a rural town in Uttar Pradesh to Kolkata in the 1970s. Yet it was clear the steady stream of customers was not there for Guru Prasad, nor for his auto parts; the focal attraction was a younger man just shy of thirty glued to a white computer monitor. Sunil Shaw was Guru Prasad's son. He and his brother, Sanjay, had started their own subbusiness—within the auto parts store—as a registered dealer for MoneyOnMobile.

"There was no future in simply peddling auto parts," Sunil Shaw told me, as I sat down beside him at his computer one afternoon in his cramped corner of the store. As he told me his story, I locked eyes with his father, who beamed at me, obviously supportive of his son's dismissal of the auto parts business. Sunil Shaw spoke to me in Hindi rather than Bengali, the traditional language of Kolkata. Most of the city's trading community was dominated by migrant laborers from Bihar and Uttar Pradesh, which were predominantly Hindi-speaking areas. "There was good money in mobile transfers, with very little investment—that's why I got into this work."

When mobile top-ups became popular in the early 2000s, small roadside stalls registered as cell phone and data top-up specialists. In these makeshift stores—*kiranawallahs*, as they're called—Airtel, Vodafone, Reliance, and other mobile operators had enlisted a small army of local salespeople. Customers would pop by asking for a few hundred minutes of talk time or a little bit of data, in exchange for which they'd pay cash to the store owner, who would then top up the customer's phone instantly via an SMS string code. It was that easy. The money would later be collected by couriers sent by the companies. MoneyOnMobile took the same concept a few steps further. Dealers like Sunil Shaw were able to process deposits to bank accounts, book train tickets, pay electricity and gas bills, and more.

As we were chatting, I got my first demonstration. An old man—a construction site worker—had come by to deposit some money. He had a bank account under his wife's name in the village, and he wanted to send her his week's savings. He handed over a small wad of notes to Sunil Shaw, who then instructed his brother to punch in an account number that was associated with the worker's mobile number. Sanjay was using bespoke MoneyOnMobile software to make the transaction. A few seconds later, the money had been sent, making its way electronically to a village in neighboring Bihar. The construction worker waited for a confirmation as he lit up a small *beedi*—a hand-rolled cigarette—from a lit coiled rope hanging ominously from the store's roof.

"So simple!" exclaimed Sunil Shaw, as he announced the transaction's completion. He had a made a cut of less than 1 percent on the transfer. A brisk stream of customers popped by in the couple of hours I sat beside him. In all, the Shaws were making somewhere between 1,500 and 3,000 rupees a day in commissions—not huge sums, but far more than they were making from the auto parts business. I did wonder, however, whether their work would survive if regular people learned how to perform these transactions themselves.

"See, you can give people smartphones all you want," declared Sunil Shaw. "But you can't make them smart."

Aviroop Roy, the Kolkata-based national sales head of MoneyOnMobile, said the Shaws were typical of MoneyOnMobile's network of dealers. "How many Indians truly know how to use the internet? How many Indians transact online? And how many are comfortable enough to trust digital money? The numbers are still very small," he assured me.

Trust certainly seemed to be in abundance among the Shaws' clientele. In several cases, Sunil recognized the customers handing him cash and was quickly able to deposit their money because he remembered their account details. More than half of the transactions he conducted were remittances to accounts in villages; some were electricity bill payments; a rare few were train ticket bookings. Using the Shaws for small errands like bill payments could in theory save people several hours that would otherwise have been spent waiting in a long queue. And while the Shaws' business saw mostly inflows of money, a similar corner shop in a typical village would see the opposite: there, people could withdraw money from their bank accounts simply by handing over their cards to the store owner. This was especially useful given the shortage of banks and ATMs in rural India.

"We are helping the government digitize the accounts of people," said Roy. "And this will keep growing. We have traditional brick-and-mortar dealers like the Shaws, but we are also talking about moving retailers in the future. Why can't a bus conductor be a *kiranawallah* one day?"

Nonetheless, Roy revealed that MoneyOnMobile's business declined by more than half in the month following demonetization, as workers stopped remitting cash to their homes. Yet by July of 2017 transactions had recovered. Clearly, a large chunk of India's population still needed assistance with payments.

Paytm or MoneyOnMobile, digital or cash, unassisted or assisted money transactions: a central question is which type of service will dominate India's future. As more Indians get smartphones—and learn how to use them—it seems logical that digital wallets will keep picking up steam. Given that its business is predicated on India's infrastructural failings and illiteracy, MoneyOnMobile will likely always take a slice of the market. If the markets were anything to go by, however, investors at least were putting their faith in digital wallets: in 2017, Paytm was valued at slightly over $6 billion, while MoneyOnMobile was still trading for pennies on the dollar in the United States.

"Cash is going to remain the dominant player in India," according to Harold Montgomery, the CEO of MoneyOnMobile. "India's not going to become Norway in one year."

Perhaps Montgomery saw something the markets had missed. From what I had seen, India was cradling versions of both Norway and a third-world economy in one basket, with both systems flourishing simultaneously in an upstairs-downstairs, haves-and-have-nots divide, forever running in parallel.

At the end of August 2017, nine months after demonetization, the country's central bank finally revealed the data everyone wanted to see: the percentage of old 500- and 1,000-rupee bills that had been returned to the banking system.

Some economists had predicted that as much as a third of the old currency would never be returned, as corrupt hoarders feared getting caught. The theory went that the unreturned money would mean lower liabilities for the central bank. Sujan Hajra, an economist at Anand Rathi Securities, told the *Wall Street Journal* that the state coffers could get a windfall of between $44 and $75 billion.[3] Finance Minister Arun Jaitley had also said as much: "Obviously, people who have used cash for crime purposes are not foolhardy enough to try and risk and bring the cash back into the system because there will be questions asked."

The theories turned out to be wrong. The Reserve Bank of India revealed that 99 percent of the old notes—about $239 billion—had been returned. Less than 1 percent of cash was unaccounted for.[4]

The other demonetization hope, that large caches of counterfeit currency would be exposed, was also proven false. The state's numbers showed that less than 0.0007 percent of the returned cash was fake currency.

There was only one conclusion to draw: demonetization had failed. If the cash recall's raison d'être was to catch hoarders of black money off guard, the move had missed its target. Perhaps this shouldn't have been surprising. In a *New York Times* essay written two weeks after demonetization, Kaushik Basu—a chief economic adviser in the previous government—had warned that most so-called black money wasn't money at all: instead it was "held in gold and silver, real estate and overseas bank accounts."[5] Those deposits remained intact.

The government had, as noted, set 250,000 rupees in old notes as the maximum amount individuals could exchange for new ones. In doing so, the hope had been that people with more cash would be forced to explain where they got so much money. India's super wealthy are nothing if not enterprising. Several reports had emerged of the rich getting employees to deposit the cash on their behalf; the mules would then repay their employers in new notes following a period of time, presumably keeping a small fee for their services. Several other laundering schemes emerged. There were reports of Hindu temples, for example, becoming fronts for laundering money, safe in the knowledge that the taxman wouldn't dare attempt a crackdown at a holy place.

Meanwhile, crises abounded in the wider market. As cash continued to be scarce, demand for basic food products began to fall. Market rates for onions, tomatoes, and potatoes declined by between one-third and one-half. Those falls in turn triggered a collapse in the agricultural sector, with farmers reporting historic levels of debt. Predictably, those developments then contributed to India's GDP growth falling to 6.1 percent in the first quarter of 2017 and an even more anemic 5.7 percent in the second quarter—down from an average 7.9 percent the previous year. Economists were beginning to piece together reports of hundreds of thousands of jobs lost.

Untold levels of pain had been wrought across India—especially for daily wage laborers and the rural unbanked. If there was one ray of light, it was the surge in the growth of digital wallets and bank accounts. "It was the right objective," admitted Basu. "India had to move in that direction. But shocking the economy with a four-hour notice was definitely not the earthquake we needed!" Electronic payments had grown 21 percent in an eight-month period since demonetization, though even those figures included a 20 percent drop since hitting a peak in the month of March. In other words, once cash began to flow freely in the market again, people became less enthusiastic about digital wallets and e-payments, reverting instead to physical money. White money would soon be converted back to black.

Given the immense trouble most citizens had endured, perhaps the greatest paradox was that the government seemed to have gained political mileage out of demonetization. In March of 2017—six months after the cash recall—Modi's Bharatiya Janata Party won a landslide three-fourths of the seats on offer in elections in Uttar Pradesh, a state with more than 200 million people. It was the greatest endorsement the ruling party could have asked for.

Shailesh Kumar of the geopolitical risk consultancy Eurasia Group told me that the win showed "Modi's efforts to tackle corruption were a far bigger draw than any negative consequence attributed to demonetization."[6]

This reminded me of Sarvesh Kumar and how he was willing to bear financial pain at the mere hint of trouble for richer, corrupt Indians.

I tried to reconnect with Kumar. I had saved his number, but he had stopped picking up his phone. He no longer lived in the apartment in Vasant Kunj. When I revisited Kumar's old auto-rickshaw stand, none of his friends seemed to know where he was. They assumed he had gone back to his village in Uttar Pradesh. I kept searching. Finally, one day, I got a lead: an auto parts dealer had a new number for Kumar.

I called. He explained that he had lost his phone, and with that all his contacts. He had also moved house. We agreed to meet for lunch the next day.

When the time came, I walked outside my office building to catch his auto-rickshaw at the curb. From a distance, I recognized his toothy smile, but he was much changed. He had shorn off his hair and lost weight. He looked malnourished.

I sat behind Kumar, who described how he had lost his phone a few months back and hadn't been able to buy a new one until recently. I spotted a pair of white earphones hanging by his shirt. His new phone had a radio on it. But it was clearly a hand-me-down: the device's body looked like a small brick, with its tiny one-inch screen displaying a crack across its face. This was no smartphone—it seemed more like a relic from the late 1990s.

We decided to go to an Indian restaurant nearby. As we pulled up, parked, and walked inside, Kumar stopped me. "I've never been inside a restaurant before," he confessed. "Will they let me in?" he asked, pointing to his shabby gray outfit, toes sticking out of his rubber sandals.

I reassured him they would.

Once inside, we ordered food. It struck me that Kumar had never opened a restaurant menu before. The roadside *dhaba*s he would frequent were much simpler, with just a few staples served. In any case, Kumar couldn't read or write. I felt overwhelmed by a surge of pity, wanting to show this young man a good time and give him a hearty lunch. I explained how people used napkins at restaurants and how the forks, spoons, and knives were placed beside the plate. I asked Kumar what he felt like eating. He smiled shyly. I recited the names of some of the dishes on offer. "It's been a while since I ate paneer," he admitted. Paneer was cottage cheese, usually prepared in cubes, immersed in vegetables and gravy. In the market, paneer had become more expensive than chicken these days. We ordered a plate of the paneer masala, a black dal, raita, rice, and some rotis. I realized this was a wild extravagance for Kumar.

"So, what happened since we last met?" I asked, as we waited for our food to arrive.

Kumar's story was heartbreaking. His optimism that his landlord would let him off a month's rent had proven misplaced. The opposite happened. Faced with a cash crunch, the landlord had raised rents by 50 percent. The justification he gave was that demonetization would force him to report his cash income, and therefore he would have to pay income tax in the future. So he needed to pass on those losses to his renters. Kumar's father was fired from his job. Father and son—and Kumar's wife, Bandhana—suddenly found themselves without a home.

"It was the biggest embarrassment of my life," said Kumar, his eyes determinedly avoiding mine. "Those one or two months were like hell. We somehow lived on the streets like beggars. I don't know how we survived the winter."

I was aghast. "And now?" I asked.

The Kumars had bounced back, somewhat. They had moved to a slum on the outskirts of Delhi, near the new satellite city of Gurgaon in Haryana. It was cheaper there. Auto-rickshaws rarely found much business in those areas. Kumar now had to drive all the way to Delhi to find his first ride. His new apartment was even smaller than the last one. They were now sharing a bathroom with two other families housed in rooms nearby. They had lost what little furniture they had. The Kumars in effect were forced to start from scratch. Asha Ram had taken a loan from a friend; it was the first time the family had ever been in debt.

As our food arrived, Kumar teared up. "I've never eaten this kind of food before!" he sputtered. We ate in silence for a while. He began to giggle as he tried to use his fork, while I tried to reassure him he could use his hands at the restaurant—it was Indian food, after all.

Kumar described how he was now trying to learn how to read basic Hindi and understand how to decipher maps. He wanted to become a driver for Ola, which had a large network of auto-rickshaws signed up to its service.

"I don't want to go through so much pain ever again," he continued, describing to me how he and his wife would eat plain rotis with salt because they couldn't afford basic lentils or vegetables. His father was still hunting for a job. "I have to figure out a way to support my family. Now that it's been a while since demonetization, cash has returned to the market and my business is back to normal. But now I need to secure my finances for the future." There were several hurdles before Kumar could drive a car for Uber or Ola, however. He would have to get a new license to drive four-wheeled cars; and he would need to save up money to purchase a smartphone and then learn how to use it.

I asked whether he still thought demonetization had worked.

"*Bilkul*," he replied. Absolutely.

But how? I asked. Surely given his pain, his family's suffering—everyone's suffering—as well as the data showing that all the so-called black money had come back, he didn't still believe it had been worth it.

"That's all fake news," he announced. While our conversation had been in Hindi, Kumar said "fake news" in English: he had picked it up from some friends. Kumar was using that phrase with a journalist, but I knew he didn't mean any malice to me in particular. "I've heard those stories. But I know that the rich have suffered. And this is just the start. Modi is going to do this again. And again. And he'll find other ways to hurt the rich and the corrupt people of India. I've heard him speak in person. I trust him, and I know he's going to come good."

I realized there was little point in arguing. Kumar had, of course, voted for Modi's party in the recent Uttar Pradesh elections. His blind faith was stronger than any data or evidence I could produce. He needed to believe in it. I wondered what would happen if Kumar one day realized demonetization hadn't been the success he thought it had.

I paid for lunch, taking care to ensure Kumar didn't see how much it had cost. We walked out, our bellies full, and hopped back into his auto-rickshaw. Kumar dropped me off at my office. We agreed to stay in touch and said our goodbyes. As he drove away, it occurred to me that Kumar simply didn't have the education or tools to survive in the kind of smartphone-first economy many were promising. The country's push to digitize—a program that people like Kumar were embracing—seemed destined to elude him. India's great digital leap forward was going to leave Kumar, and millions like him, behind.

Conclusion

The Everything Device

THE WORLD CHANGED on January 9, 2007.

It was the Macworld trade show in San Francisco, an annual showcase for Apple products, and founder Steve Jobs was about to introduce a new gadget.

"Every once in a while, a revolutionary product comes along that changes everything," announced Jobs.[1] The Macworld audience had a Pavlovian expectation for something game-changing that day. In 1984 Apple introduced the Macintosh, which went on to transform computing and make the mouse a mainstream accessory. Then in 2001, the iPod arrived. "It didn't just change the way we all listened to music. It changed the entire music industry," Jobs reminded his audience. (This was no exaggeration. When Apple began offering individual songs for ninety-nine cents on its iTunes store, the era of record companies selling entire albums was shattered.)

"Today, we're introducing *three* revolutionary products. The first one is a wide-screen iPod with touch controls."

Jobs paused for dramatic effect.

On cue, the audience broke into hearty applause.

"The second," continued Jobs, "is a revolutionary mobile phone."

This time, before he could pause, cheers rang out—with a louder, prolonged burst of clapping. Apple had never manufactured a phone before.

"And the third," he went on, as a big screen behind him mirrored his words, "is a breakthrough internet communications device."

A whoop, followed by a polite round of clapping; by now the audience was a bit confused at the deluge of new products.

Jobs let his words hang in the air, teasing the crowd as it waited in anticipation.

"So, three things," he recapped, as the screen behind him showed three Apple icons representing an iPod, a phone, and the internet. "A wide-screen iPod with touch controls, a revolutionary mobile phone, and a breakthrough internet communications device."

Silence.

"An iPod, a phone, and an internet communicator," Jobs repeated, as the screen displayed each of those icons in the center, flipping to reveal the next one. The animations behind Jobs had been carefully choreographed to match his words.

Slower and louder, he tried again: "An iPod...a phone...are you getting it?" teased Jobs, ever the showman, as a few people laughed nervously, beginning to cotton on. The punch line: "These are not three separate devices. This is *one device*...and we are calling it *iPhone*."

This time the Macworld delegates brought the house down.

While so-called smartphones were already in existence—some had the internet, some had cameras, some had touchscreens—nothing that came before tied everything together in a way that made it all look so simple, obvious, and cool.

The iPhone would become, as Jobs promised, a "revolutionary device." Sales would surpass even the most bullish expectations. As Brian Merchant described in *The One Device*, a book about the iPhone's history, no other retail product has ever sold as many units. By 2016, Toyota had sold 43 million Corolla cars, Sony 382 million PlayStations, and J. K. Rowling 450 million Harry Potter books. At 1 billion units and counting, Apple's iPhone had outstripped them all—combined. The iPhone was a winner in every category: it was the world's bestselling camera, video screen, computer, phone, music player, map, alarm clock, calculator, and much more. It was also, as Merchant reported, the world's most profitable product, well ahead of the second-place Marlboro cigarettes. By 2017, Apple had grown to a market capitalization of $900 billion, a valuation greater than the entire GDP of Turkey or Argentina. This was largely thanks to the success of one device, which had racked up well over a trillion dollars in revenue.

India may be the one large democracy Apple hasn't been able to dominate. With new phones priced as high as $1,700, it would cost most people a year's salary to buy one.[2] As a result, Apple currently makes less than 1 percent of its global revenue from India. (The company is now trying to sell older, refurbished phones in a bid to improve its sales.)

Even with a small footprint, however, the iPhone has had an outsized impact in India. By reimagining what was possible in one single device, and by making it stylish, desirable, and user-friendly, Apple helped forge a path for Asian companies like Xiaomi, Samsung, and Micromax.

Consider again the analogy with cars that began this book. A century ago, Henry Ford asked himself what kind of car ordinary Americans would buy. He imagined a vehicle that was cheap and easy to maintain, with clearance high enough for use near farms and areas without proper roads. It was a populist vision. Ford created the Model T: an automobile that was priced at just $850, about half the price of its competitors. The Model T was a game changer. Americans began to believe that cars were for everyone; they began to understand what this could mean for their families, communities, and businesses. It sold 15 million units, spurring a boom and making cars mainstream.

Asian smartphones are India's Model Ts. They are not only cheap but are also rugged and designed for local conditions. Some prioritize battery strength because the supply of electricity can be spotty; others allow for not one but two SIM cards because cellular service is erratic. Almost all of these phones allow for side-loading, because that remains how many middle-class Indians exchange videos and content.

Steve Jobs called the iPhone "revolutionary" because it pioneered the idea of combining several high-quality gadgets in one device. Yet each of those products individually—portable music players, phones, cameras, computers—was already in widespread use in the West. In 2000, only 6.4 million Indians—or 0.6 percent of the population—had mobile phones. Two percent of the population had access to internet-connected

PCs. A similar tiny fraction of Indians had cameras or portable music players. And this has been the argument at the heart of this book: if smartphones are "revolutionary" for the West, they are having an even greater impact in places like India. Cheap smartphones have not only combined several gadgets into one device but have also *introduced* those devices to Indians. Smartphones are bringing computing, photography, and video to India's middle classes. The ramifications of this mass proliferation of technology likely goes well beyond what even Steve Jobs imagined.

And yet, leaving aside its financial success, the iPhone's legacy feels mixed at this moment—at least in the West. In 2007, it was widely heralded. By 2017, the mood had changed. Farhad Manjoo, in his popular State of the Art column for the *New York Times*, wrote of how devices like the iPhone now came "freighted with worry."[3] He added that while "tech might improve everything . . . it's probably also terrible in ways we're only just starting to understand." In January 2018, two leading investors in Apple—JANA Partners LLC and the California State Teachers' Retirement System—publicly expressed concern that the entrancing qualities of the iPhone could pose a mental health danger to consumers, and especially children.[4]

As outlined in Chapter 6, the worries stem from several sources now: smartphone addiction, psychological damage, negative effects on cognition and well-being, fake news, the security of our data, the role of Russian hackers in influencing the 2016 US presidential election, and more. A growing consensus seems to be building in the West that we have entered a dangerous era of constant connectivity, one where people can be influenced and controlled, one where everyone seems somehow less fulfilled and unhappier than before. The popular Netflix series *Black Mirror*—the title is a play on the smartphone screen—depicts scenarios where technology has taken control of human life. Unlike similar futuristic series from the 1970s and '80s, however, the stories in *Black Mirror* often feel completely plausible—and not just science fiction.

The warning signs are useful. But perhaps the pendulum has swung too far the other way. As the technology historian Melvin Kranzberg said in 1985, "Technology is neither good nor bad; nor is it neutral."

For the West, India presents some perspective. The debate there is not as "freighted" with fear, partly because the smartphone is perceived to be an equalizer, something that can help the hundreds of millions of Indians who lack proper access to education, health care, and work opportunities. The mood is also different because technology has not yet reached anywhere near saturation point—half the population is still offline. This could yet be an advantage. India can use this moment to observe the West's soul searching and prepare its consumers.

India's journey to an age of constant connectivity is an unlikely one—especially when you consider that in 1980, for example, it had just 2.5 million telephones to serve 700 million people, or 1 phone for every 280 people.[5] There were only 12,000 public telephones across the country. In general, the quality of service was atrocious. Breakdowns were common. Requesting the telephone department to fix a problem usually entailed paying a bribe. Even so, the list of Indians waiting to get a telephone line grew into the millions. But there was little urgency on the part of India's government. Once, when a politician complained in Parliament about the sorry state of telephony, he was publicly rebuked by Communications Minister C. M. Stephen.[6] Telephones were a luxury and not a right, Stephen said, adding that any Indian who was not satisfied with the service could return his device so it could be passed on to one of the millions waiting for a connection. Stephen's attitude was hardly unique; his prime minister, Indira Gandhi, was similarly known to dismiss the importance of technology in a country where farmers needed simpler things like food, water, and an education.

As the writer Gurcharan Das describes in *India Unbound,* an Indian American engineer named Sam Pitroda had a different point of view. Pitroda grew up poor in rural India but had become a telecom millionaire in the United States. He believed it was a mistake to think India should wait to become wealthy before telephony became commonplace. Instead, according to Pitroda, India needed telephones to prosper, to unleash efficiency and productivity. On one of his visits to New Delhi, he set up a meeting with Gandhi and her son Rajiv, who would succeed her as prime minister in 1984. Pitroda pitched a plan to put India on a path toward universal telephone access. The Gandhis agreed and Pitroda set to work. He is now credited with creating a low-cost, indigenous digital network, building fiber optics to connect Indian cities and districts. The number of telephone users grew dramatically: there were 20 million telephones in service by the mid-1990s. This, of course, was still just a fraction of what needed to be done, which Pitroda believed was the equivalent of a telephonic New Deal.

Despite the growth, there remained several hurdles to connecting the country's poor. Pitroda came up with an idea for an Indian-style phone booth. It was well known that coin-operated public phones in India rarely worked. Intense heat and humidity often broke them down; they were badly maintained. Pitroda felt that if he could put meters on regular telephones and give them to shopkeepers in busy marketplaces, it would unleash telephone usage for the masses. The theory proved correct. The operators made a commission on each metered call, and because they wanted to stay competitive, they ensured that their devices were properly maintained. The system of PCOs, or public call offices, grew exponentially. By 2000, PCOs connected India in a way the private telephone network never could: there were half a million outlets across the country's bazaars, connecting—in theory—hundreds of millions of users. (The boom in PCOs also created more than a million jobs. It became common for mom-and-pop-style entrepreneurs to set up a PCO booth in their stores as an additional source of income.)

Telephony was still run entirely by the government. No matter how visionary, the public sector remained ill equipped to innovate and create the competition necessary for faster growth.[7] When the Indian economy began to open up, the Department of Telecommunications decided to hold auctions for licenses to operate mobile services. The process turned out to be slow and corrupt. The government gave state-owned operators a cellular license for free while charging huge sums to private sector carriers, thereby putting them at a disadvantage.[8] The charade ended up causing heavy losses for companies, needlessly delaying an aggressive rollout of cellular services across the country. It wasn't until the mid-2000s that private operators finally overtook their state-run competitors, attracting customers with sustained spending on advertising and outreach.

A major setback emerged in 2011. A government auditor accused Andimuthu Raja, the telecom minister at the time, of being responsible for the loss of $30 billion in revenue. The auditor's report found that

Raja had sold 2G spectrum licenses at a fraction of their real value—ostensibly to receive kickbacks. The "2G spectrum scam," as it came to be known, became one of the biggest corruption scandals in Indian history. (Note: In December 2017, a special court threw out the charges, saying the prosecution had "miserably failed to prove its case." Raja was acquitted.)

Had India cut itself off from the rest of the world in 2011, right after the discovery of the alleged 2G scandal, none of the stories in this book would have taken place. India's mobile sector could easily have gone the way of some of its other creaking state-run services. But in 2012, India welcomed 3G mobile technology. Finally, a service had arrived that made internet browsing on mobiles relatively fast.

"3G was our inflection point," says Vineet Taneja, a former CEO of the Indian mobile manufacturer Micromax. Taneja has also worked in senior positions at Nokia, Samsung, and Airtel. "We had just begun to get phones, but we didn't have connectivity. Suddenly, we all sat up and went 'Wow—this is going to change everything for India.'"

Around the same time, the Android operating system had become popular. "It was no coincidence," argues Taneja. "Google was watching India closely. They wanted to ride the bandwagon here. They realized what India was: let alone mobile first, we were a mobile-only opportunity. This was the only way in."

Manufacturers like Samsung and Taneja's Micromax cashed in, flooding the market with large-screen smartphones that could serve as internet browsing devices. A flurry of private sector investments began. Companies were developing multilingual software to reach a greater number of Indians. E-commerce took off, further fueling a race to get online. Indian start-ups became fashionable. Other Chinese and Indian smartphone companies joined the fray. Everything seemed to be driven by the potential of internet-enabled smartphones.

"It was like pouring water in the desert sand," remembers Taneja. "Whatever you did, the market would absorb." Indians were willing to invest great amounts of their savings in buying smartphones. At Micromax, for example, the bestselling phones were never the cheapest ones. A $100 phone would sell better than the cheapest $80 device, says Taneja. "Indians began to understand the value of how these things could change their life. They were investing in their futures."

By 2015, millions of Indians were coming online every month, and 4G technology arrived. Mobile internet was truly becoming a replacement for PC-based browsing. People could purchase 4G dongles to set up Wi-Fi networks at their homes or offices. Companies were engaged in cutthroat price wars to grow their footprints. The greatest disruptor here was undoubtedly Reliance's new Jio service, as noted earlier. The group's petrochemicals and refinery businesses were bringing in huge amounts of near-guaranteed revenue every year. For Reliance chairman Mukesh Ambani—India's richest man—cellular internet seemed to be the business with near unlimited potential, and he was willing to take a short-term financial hit to capture the market. When Jio launched in September 2016, the 4G network was offered to Indians for free through the end of the year. A scramble to sign up began. "Life is going digital," Ambani told investors at his company's annual general meeting that year.[9] In December, Ambani extended the free internet offer for a further three months. More than half a million Indians were signing up for a Jio SIM card every day. Jio's aggressive entry meant other players would suffer. Aircel, the country's fifth-largest wireless operator, was forced to declare bankruptcy in February 2018; meanwhile Vodafone's Indian operation merged with Idea Cellular, and Bharti Airtel partnered with Norway's Telenor.[10] The sheer number of free 4G internet users meant that national bandwidth had gotten clogged. OpenSignal, the wireless-coverage mapping company, found India ranking seventy-fourth of seventy-five countries surveyed for 4G speeds.[11] But even with the inconvenience, more and more Indians kept coming online. More users meant more consumption, more opportunities. The pie would keep growing.

"The best thing the Indian government did was to just get out of the way," maintains Taneja. "In hindsight, that was the single biggest tipping point of them all. It happened in the nick of time. Otherwise we wouldn't be where we are right now."

In several industries—coal and railways, for example—the government had sat idly by as state-run enterprises stifled growth and innovation. But in the telecom sector, even with corruption and roadblocks, the private sector had truly taken off and transformed connectivity. This was India's unintended New Deal.

The idea for this book came in 2014. I had just moved to New Delhi for a new job as CNN's South Asia bureau chief. For my first few weeks there, I was put up in a hotel room. At night, I would try to get a sense of the national mood by watching TV. There were seemingly endless numbers of channels—sports, news, movies, soap operas, dramas, and talk shows. Yet what struck me most of all was not the content itself but the advertisements. Indian commercials tend to be narrative driven, recounting everyday stories. I have often thought these commercials get at the heart of what middle-class Indians are feeling. It took me only a few evenings of browsing to realize that the most prominent advertisers were the ones selling either mobile devices or cellular data. Their ads were omnipresent on billboards, buses, and corner shops, but it was on TV that they resonated, selling an aspirational notion of what it meant to be connected and empowered.

One commercial in particular stood out. It was part of a campaign for the cellular provider Idea (which later merged with Vodafone). Early in 2014, hundreds of millions of Indians were readying to vote in the biggest election in history. Idea captured the moment with an inventive commercial. The scene was set in a stereotypical Indian village, sleepy and rustic. A politician was making his campaign stump speech in Hindi. "I'll give you running water," he declared. "I'll deliver electricity, twenty-four hours a day!" With each soaring promise he became more animated. "I'll give you roads! *Bridges!*" Nearby a couple of young men were sitting on a fence. One of them was immersed in his smartphone. "Aha, not so fast, Mr. Politician," he called out. "Look here," he said, waving his phone as it played out a video. "You made these same promises at the last election too. We might be villagers, you know, but we're not fools!" The politician looked suitably mortified; the villagers began throwing trash at the disgraced candidate; a catchy Hindi jingle broke out: "We have the power of the internet, no making-a-fool-of-us, no making-a-fool-of-us."

The commercial was an exaggeration, of course, but it mined several trends simultaneously playing out across India. It showcased how the internet was promising to reshape a playing field that had been drawn on boundaries of caste, tribe, gender, power, money, geography, and language. It told the story of the rise of India's poor and lower-middle classes, packed—yet forgotten—in small towns and villages across the country. It reflected the rising power and impudence of India's youth. It was about New India, the India of the internet era.

Idea ran several other ad vignettes with similar "don't make a fool of us" themes. (In Hindi, the jingle's chorus was "*no Ullu banao-ing*," Hinglish for "don't make me an 'owl'"—a fool.)

One commercial showed a fraudulent *baba*, or spiritual guru, who declared there would be plentiful rain if he was given gifts of gold. However, even as unwitting Indians queued up to offer their jewels, a bespectacled schoolboy interrupted the proceedings, displaying the weather app on his phone for all to see: The forecast was for rain, he pointed out, so why pay up in gold? Cue the jingle.

Another vignette showed an auto-rickshaw driver winding through busy traffic. Along the way, he turned around to his passenger and, with an innocent smile, informed him that he was taking a "special shortcut." But the passenger was tracking the journey on his smartphone, and his map application showed otherwise. "No, no, it's a 'longcut.' Who do you think you're trying to fool?" he cried out. Again, cue the jingle.

Yet another Idea ad showed a white man in a suit, speaking in an exaggerated British accent. He was addressing an auditorium of Indian students and telling them to study at his university; he claimed he had a track record of placing graduates in top jobs in the United Kingdom and the United States. His audience seemed impressed. But one student Googled the name of the university in question. It turned out the school wasn't accredited in either England or America. "Don't make fools of us!" admonished the student loudly. "Your college won't get us jobs anywhere. Look, I checked. We have the internet here too, you know?"

And so on. What struck me about the ads was how they captured everyday situations. People were conning each other, looking to make a quick buck here and there in a low-trust society with entrenched corruption, dramatic inequalities, and a shortage of opportunities. What Idea was offering was a solution: the internet-enabled smartphone. That was my starting point.

In Part One of this book, I focused on opportunities. Phoolwati's life is very different with a smartphone in her hand. She is a fearless leader, a role model in her village. Her story runs in sharp contrast to those of the nameless and numberless girls who were denied the right to own a smartphone, unable to escape from the horrors of village life. Phoolwati's story is encouraging, but it is also one of chance. More than any other place, India is a lottery.

In Abdul's story the smartphone helped him realize his dream of becoming a teacher-motivator—quirky though his coaching institute may be. The phone also turns out to be a hopeful development in early childhood learning, an area in which India has long lagged behind the rest of the world.

Sudhir, Renu, and Babloo, in Chapter 3, showcase how the phone can lead to different kinds of work opportunities, but with varying degrees of success. If the gig economy and the e-commerce pie grow to the extent being predicted, perhaps workers in the future will have easier lives. A comparison with other countries is sobering, however. India's e-commerce market was worth $21 billion in 2016. China is capable of topping that figure in twenty-four hours of sales. On November 11, 2017—"Singles' Day," as it's called—Chinese consumers spent $25.3 billion on online shopping.[12] India has a very long way to go.

Simran in Part Two is happily married—she says—because of the smartphone. But I wonder if she and her family were ready for the turmoil created by internet dating. Perhaps it will take a generational change before Indians truly become comfortable with dating apps and the freedoms they represent.

In my research for the chapter on pornography, I was surprised to learn just how much adult content Indians were consuming. Very soon, India will become the largest global market for porn, and this in turn will have an impact on the types of content being created. I'm reluctant to draw too many conclusions from this increased consumption—I don't want to pass moral judgment—but "rape videos" have indeed become common in parts of India, and it's hard not to see a connection.

Later, Saikat Sinha introduced me to the coming problems of his generation. Smartphone addiction is barely a topic of conversation in India, but no doubt it soon will be. Meanwhile, as Indian consumers spend more, large companies will fight harder to reach them with their products and their advertising. I hope Indian lawmakers—and civil society leaders—are watching the West's deliberations on these matters closely and considering regulations that would protect local consumers.

In Part Three, on the role of the state, the fact that smart young teens like Zeyan Shafiq look around and see no opportunities is disappointing. When I was based in Delhi I was often surprised at how little attention Kashmir's internet shutdowns received. If the government closed service in Delhi or Mumbai or Bengaluru, there would be a national outcry. But in Kashmir and Darjeeling, the authorities can cite security concerns and get away with it. It sets a dangerous precedent.

The national debate on cyber security and privacy—involving Aadhaar, especially—is polarized. New Delhi has seemingly turned a blind eye to increasingly clear flaws in its systems; activists have responded with conspiracy theories that detract from the issue itself. What mustn't be forgotten is this: Aadhaar, if channeled properly, has immense potential for India. How this debate is resolved will be a crucial test of the strength of India's civil society and public institutions. Perhaps activists can draw strength and inspiration from within: after all, their efforts led to India banning Facebook's Free Basics, thereby preserving net neutrality.

Chapter 8 explores the rise of digital money and the important ramifications it is having for business and tax collection in India. But the story of MoneyOnMobile—and its "assisted model"—is proof that many will be left behind. Sarvesh Kumar seemed so idealistic about a clampdown on corruption, yet it was clear to me that India had let him down. I have stayed in touch with him in part because I'm curious if he's ever going to find a way to learn to use a smartphone. When we last spoke he told me he is going to become a father soon.

The dozen-odd profiles in this book are only a small sample of how the world's biggest democracy is changing. There's little science in trying to use the stories to explain the churning that is under way. Phoolwati, Sarvesh, and Abdul offer a start.

The question, then, is where the transformation is taking India. At the end of 2017, my wife and I visited the beautiful island of Havelock, part of the Andaman Islands in the Indian Ocean. On arriving at our hotel, we discovered that it had no internet; we soon became aware that we had signed up for an enforced digital detox. There was one café with internet, a twenty-minute drive away, but the service there was so slow one could barely download email. A few days into our holiday we realized how calming it was to be disconnected. There were no distractions.

We read books; we had uninterrupted conversations. When the time came to leave, we were almost saddened to learn that Airtel was expected to launch 4G services across the island. The final frontier of Indian territory will soon be online.

The rest of the country, meanwhile, will jump to the new world of 5G and its blindingly fast download speeds. India's internet will welcome non-English speakers, creating more services and content in local languages. Streaming video will take off; consumption patterns will change; businesses will conduct more and more of their work online. For much of independent India's existence, the middle classes have trailed their global counterparts on adopting new technologies. Thanks to cheap smartphones, a new generation of Indians will grow up with access to technology virtually on par with that of young Germans or Americans. This is where the change will be greatest.

It is easy to be cynical about the impact of the smartphone in India. There will be problems with addiction, with cybercrime, with trolling. Many families and communities will experience disruptions in their lives; a great number will be left behind. The state could misuse technology; and the current high level of media-driven optimism will undoubtedly be followed by a period of disappointment. But perhaps this is true of any technology in history. The solution is not for India to withdraw but for it to learn from the West's missteps.

Some predictions are easier to make because the trends are already clear.[13] By 2024, demographers expect, India will overtake China as the world's most populous country. India will retain this position for the rest of the century, peaking at a population of 1.65 billion by 2060.[14] With that many inhabitants, India's progress will determine a large part of the world economy's health in the twenty-first century.

India will soon also be the world's youngest country, with a median age of around twenty-five. I touched on the "demographic dividend" debate in Chapter 3, but it is difficult to stress enough that India's youth represent both the country's greatest hope and its greatest challenge. In the 2020s, the 700 million Indians under the age of twenty-five are going to be a force to reckon with. They will be digital natives—connected, hopeful, and empowered—and they will be smarter than their politicians in a country where most leaders are over the age of sixty. Who will lead them, and how, is a worrisome question. This young generation will need gainful employment and will begin to demand things their parents and grandparents did not: a work-life balance, holidays, entertainment, dining out (eating meat and drinking alcohol), independence, privacy. The smartphone will enable—and encourage—some of these new desires. But it will also lead to disappointment for many. Technology will never fix poverty, inequality, and broken infrastructure.

The flip side of a youthful India is a topic that doesn't get much attention. India will soon have hundreds of millions of senior citizens. In 1960, the average life expectancy for an Indian was just forty-one years; today, one can reasonably expect to live to seventy.[15] In the 2020s, India will begin to grapple with more young people leaving home to find work, with older parents finding themselves alone and unprepared for their retirement. As the system of large families living together disintegrates, India will need to find alternatives for its senior citizens, from old-age homes to pensions and health care services. There is little sign of a plan to deal with this.

Two other trends are locked into India's future. According to the last census, the country has 943 women for every 1,000 men, a statistic indicative of how families have systematically aborted female fetuses. The problem will get worse. There are only 919 baby girls in India for every 1,000 baby boys. India's future looks testosterone heavy, and there will soon be a shortage of women for young Indian men to date and marry.

Religion will continue to shape India's future and its debates. By 2050, India will have the largest Muslim population in the world, surpassing Indonesia and ahead of Pakistan and Bangladesh.[16] As India continues to become more polarized on sectarian issues, a confrontation between the world's largest Hindu and Muslim populations is a realistic worry.

The smartphone will not play a defining role in the trends mentioned above. Instead, it will serve as a catalyst of sorts, speeding up changes already under way. It will help provide more data on how people are getting educated, working, and living their lives. This information represents an immense potential force, and how the government and large corporations use it will help determine India's—and therefore the world's—future.

India Connected is a story about change—but it is a story that has just begun. The next chapters will depend largely on how technologies like the smartphone are harnessed and regulated. And the future is unlike centuries of the past; India can now control its own destiny. A new generation of connected Indians has an immense opportunity to act responsibly in shaping the country in which they want to live. Abdul once told me something—he may have picked it up on YouTube somewhere—that I have taken to heart. "There are a million things that can go wrong in India," he said. "But there are a billion reasons to keep trying."

ACKNOWLEDGMENTS

This book is not mine alone. I owe an immense debt to the many people who have allowed me into their lives and told me their stories. Several of them are named and described in this book, but there are also many others who have given me generous amounts of their time. Each of their stories has shaped the course of this book. I can only hope I have done them justice.

It is humbling to remind myself how much help I have had along the way. Zoë Pagnamenta, my agent in New York, made sure *India Connected* saw the light of day. Zoë was my first champion, and she remained an immense source of support and advice throughout. Thanks also to Alison Lewis at the Zoë Pagnamenta Agency.

One could ask for no better editor than Tim Bent at Oxford University Press. Tim's guidance, patience, and wit were invaluable. His love for India was evident and heartening. India Cooper copyedited with great care, Janet Foxman and Mariah White helped with production. Any remaining errors are all mine. Thanks also to Sugata Ghosh and his excellent team in India, and Niko Pfund for welcoming me into the fold with a grand breakfast in New Delhi.

Sachin Arya in Bengaluru worked diligently across distance and time zones to help me with research and ideas.

At CNN—my employers for eleven great years—I have many colleagues to thank. Fareed Zakaria, for guiding, teaching, inspiring, and many great meals. Ellana Lee in Hong Kong, for being an amazing manager and for letting me travel and write while keeping my day job. Few other companies would allow this much freedom. Becky Anderson, Deborah Rayner, Tom Goldstone, Tony Maddox, Jeff Zucker, Andrew Morse, Rick Davis, Ram Ramgopal, Jamie Donald, and Gill Penlington all played vital roles in my time at CNN. I am also grateful for having an all-star team of journalists while I was in Delhi: Omar Khan, Sugam Pokharel, Sanjiv Talreja, Rajesh Mishra, Vijay Bedi, Ajay Bedi, Deepak Rao, S. Gopal, Rishi Iyengar, Charles Riley, Manveena Suri, Huizhong Wu, Mukhtar Ahmed, Iqbal Athas, Farid Ahmed. Their brilliance made my work easier. They ensured much of my time in the office was filled with laughter—and chocolate cake. Thanks also to the wider CNN family in the region, who gave me the space and support to write: Roger Clark, Mitra Mobasherat, Licia Yee, Bex Wright, Elizabeth Joseph, Marc Lourdes, Paul Armstrong, Hilary Whiteman, Gregory Ho, Fabrizio Palombi, Daniel Symmonds, Siddharth Jain, Anuradha Agnihotri, Deepa Sridhar, Sheena Chopra, Sonali Chatterjee, Sunita Rajan.

Several friends, colleagues, and advisers have been ever-present sounding boards, without whom this book would undoubtedly be poorer. Amitav Ghosh was my first writing teacher and has long been a great supporter and mentor, as has his incredible wife Deborah Baker. My wife and I will always be grateful for their companionship in Goa, and for the fun times with them and Lucarno, Rahul, and Aradhana. Special mention to Cecil S. Pinto for the ready advice on Chapter 5. Kunal Agarwal and Aashti Bhartia somehow found ways to dispel the gloom of Delhi's pollution in the winters. Rajiv and Janis Chaudhury and their wonderful team made living in Nizamuddin feel like a treat. Surender Kumar and Meera were our rocks. I drew sustenance and ideas from friends in India and around the world: Anand Giridharadas, Priya Parker, Mallika Kapur, Nikhil Kumar, Anant Nath, Aditya Tripathi, Amelia Lester, Michael Robinson, Sage Mehta, Hassan Sultan, Ali Sethi, Amar Bakshi, MacKenzie Sigalos, Caleb Franklin, Anisha Dixit, Michael Soto, David Diaz, Hytham Ashuraey, Sarah Gumbley, Siddhartha Das, Jayshree Bajoria, Gitika Srivastava, Naresh Ramarajan, Richie Banerji, Siddhartha Sinha. I am thankful to a small army of economists, writers, journalists, and technology experts for free advice: Kaushik Basu, Ramachandra Guha, Mukul Kesavan, Bobby Ghosh, Sharmila Sen, Mihir Sharma, Snigdha Poonam, Ira Trivedi, Parul Abrol, Bashaarat Masood, Pankaj Yadav, Vineet Taneja, Anirudh Suri, Osama Manzar, Rajan Anandan, Bejul Somaia, Nandan Nilekani, Kunal Bahl, Sachin Bhatia, Nikhil Pahwa, Krishna Prasad, Juli Q. Huang. And I owe a great debt of thanks to Shailesh Kumar, Milan Vaishnav, Mallika Kapur, Sania Farooqi, Rishi Iyengar, Nikhil Kumar, S. Prasannarajan, and Lindsay Ford for patiently reading drafts at various stages and giving feedback.

Writing and editing can sometimes take a toll; I have realized that the greatest burden tends to fall on the families of writers, who must suffer through mood swings, forgotten birthdays, and long absences. I'll never be able to fully thank my family for their constant, reassuring support. I have dedicated this book to my parents O. P. and Neema. In many ways these chapters are a by-product of their decision to move our family from London to Kolkata all those years ago. Fittingly, I wrote the last bits of this book cocooned in a small study on the terrace of their Kolkata home. My sisters have been my greatest cheerleaders. In San Francisco, Alka has always been my first reader and adviser, often closely followed by her husband, Ravin. Playing squash with my niece and nephew, Saffron and Rohan, was a splash of fun as I turned in the last chapters. In Singapore, Anita created a home away from home for my wife and me. Her husband, Puneet, and their boys, Ayaan and Viraj, were terrific distractions. Halfway around the world, in Pasadena, my in-laws, Jeannie and Peter, were a constant source of encouragement and advice from afar. Harry, Cris, Annie, and Sam—plus Mr. Bates and Darby, of course—kept me in good cheer.

And finally, none of this would have been possible without my wonderful wife. Emma knew I would write this book before I did. She has *lived* these chapters with me, first by selflessly agreeing to transplant our newly married lives from New York to New Delhi, and then by being the first filter for my every thought. Many of the ideas in this book have been born out of our debates about India, the media, or technology. Somehow, as we juggled our jobs and seeing each other, with more than three hundred flights in four years, we made it work. Without Emma, this journey would not have begun, or been worth it.

NOTES

Introduction

1. Several sources track India's internet users, and they vary in their findings and projections. I've chosen to use data showing people who have used the internet at least once. Some other surveys, such as Mary Meeker's annual Internet Trends report, tend to look at "active" users, which represents a lower figure. Indian government data, http://www.trai.gov.in/about-us/annual-reports; Mary Meeker/Kleiner Perkins Caufield Byers 2017 Internet Trends, http://www.kpcb.com/internet-trends; compiled data, https://www.statista.com/statistics/265153/number-of-internet-users-in-the-asia-pacific-region; "The Future of Internet" Report by NASSCOM and Akamai, http://www.nasscom.in/sites/default/files/media_pdf/nasscom_akmai_technologies_report_showcase_how_internet_changing_india.pdf.
2. Again, there are varying estimates for the percentage of women compared to men on the internet in India. The most widely cited statistic is 29 percent, as shown in "Children in a Digital World"—a 2017 report by UNICEF, https://www.unicef.org/publications/files/SOWC_2017_ENG_WEB.pdf.
3. For the purposes of comparing income levels I've used the World Bank's 2016 data for GDP per capita in current US dollars: https://data.worldbank.org/indicator/NY.GDP.PCAP.CD?locations=IN-CN-MX-US.
4. India's government conducts a national census every ten years. The last was in 2011: http://censusindia.gov.in/2011-prov-results/data_files/india/Final_PPT_2011_chapter6.pdf.
5. I chose to use GDP per capita in real terms rather than PPP (purchasing power parity) as a gauge of spending power. One reason is that products like cellular phones—unlike, say, cups of tea or coffee—often have similar prices

across global markets. Indeed, new models of the iPhone are more expensive in India, Indonesia, and Thailand than they are in richer countries like Japan or the United States—the opposite of what purchasing power economics usually argues.

6. Francis Fukuyama, *Trust: The Social Virtues and the Creation of Prosperity* (New York: Simon & Schuster, 1996).
7. Basu first recounted this story to me while teaching an undergraduate economics class at Harvard University. His argument has also been published in an academic study: K. Basu and J. Weibull, "Punctuality: A Cultural Trait as Equilibrium," in *Economics for an Imperfect World: Essays in Honor of Joseph Stiglitz*, ed. R. Arnott, B. Greenwald, R. Kanbur, and B. Nalebuff (Cambridge, MA: MIT Press, 2003).
8. Gurcharan Das, *India Grows at Night* (New Delhi: Penguin India, 2013).
9. Interestingly, you will often find Indians storing their credit cards in the same white paper sleeves the bank would deliver them in. The sleeve preserves the card, but it is also a symbol of how infrequently the card is pressed into service.
10. As told to me by Uber India president Amit Jain in an in-person interview in April 2017. After India's cash recall in November 2016—"demonetization"—cash transactions fell steeply, but they recovered once the supply of notes returned to the system.
11. The Global Competitiveness Report 2017–2018 by the World Economic Forum, http://www3.weforum.org/docs/GCR2017-2018/05FullReport/TheGlobalCompetitivenessReport2017%E2%80%932018.pdf.
12. Corruption Perceptions Index 2017 by Transparency International, https://www.transparency.org/country/IND.
13. The Global Gender Gap Report 2017 by the World Economic Forum, http://www3.weforum.org/docs/WEF_GGGR_2017.pdf.
14. "Crime in India 2016," National Crime Records Bureau, Ministry of Home Affairs, http://ncrb.gov.in/StatPublications/CII/CII2016/pdfs/NEWPDFs/Crime%20in%20India%20-%202016%20Complete%20PDF%20291117.pdf.
15. "Me, Myself, and My Killfie: Characterizing and Preventing Selfie Deaths," http://precog.iiitd.edu.in/blog/2016/11/13/selfie-killfie-deaths/.
16. "2017 Year in Review by Pornhub Insights," https://www.pornhub.com/insights/2017-year-in-review.
17. Karl Marx, "The Future Results of British Rule in India," *New York Daily Tribune*, August 8, 1853.
18. I've used gross national income per capita in current US dollars for this calculation, per World Bank data: https://data.worldbank.org/indicator/NY.GNP.PCAP.CD?locations=IN.
19. Tim Cook's comments on India were made during a call with investors on August 1, 2017: https://www.apple.com/newsroom/2017/08/apple-reports-third-quarter-results/.
20. Jethro Mullen, "Amazon Is Pumping an Extra $3 Billion into India," CNNMoney.com, June 8, 2016, http://money.cnn.com/2016/06/08/technology/amazon-jeff-bezos-india/index.html.

21. Simon Kemp, "India Overtakes the USA to Become Facebook's #1 Country," TheNextWeb.com, July 2017.
22. Venkat Ananth and Patanjali Pahwa, "India's Internet Vision. Good Idea," The Ken, February 23, 2018, https://the-ken.com/story/india-internet-vision/.
23. Hinduism, Buddhism, Jainism, and Sikhism are generally seen as the four main religions with origins in India.
24. As described in V. S. Naipaul, *India: A Wounded Civilization* (New York: Vintage Books, 2003).

Chapter 1

1. *State of Literacy 2011* (New Delhi: Office of the Registrar General and Census Commissioner, Ministry of Home Affairs, India, 2011), chap. 6, http://censusindia.gov.in/2011-prov-results/data_files/india/Final_PPT_2011_chapter6.pdf.
2. Press Trust of India report published in the *Times of India*, February 22, 2016, https://timesofindia.indiatimes.com/city/ahmedabad/Gujarat-village-bans-mobile-phone-for-school-girls/articleshow/51087691.cms.
3. Despite rapid urbanization, India's 2011 census showed that 833 million of the country's 1.21 billion people—about 70 percent of the population—lived in rural areas.
4. *Crime in India, 2016* (New Delhi: National Crime Records Bureau, Ministry of Home Affairs, 2016), http://ncrb.gov.in/.
5. *The Global Gender Gap Report, 2017* (Geneva: World Economic Forum, 2017), https://www.weforum.org/reports/the-global-gender-gap-report-2017.

Chapter 2

1. "Macaulay's Children," *Economist*, October 28, 2004, http://www.economist.com/node/3338436.
2. *State of Literacy 2011* (New Delhi: Office of the Registrar General and Census Commissioner, Ministry of Home Affairs, India, 2011), chap. 6, http://censusindia.gov.in/2011-prov-results/data_files/india/Final_PPT_2011_chapter6.pdf.
3. *The Global Competitiveness Report 2017–18* (Geneva: World Economic Forum, 2017). Last accessed February 9, 2018, http://www3.weforum.org/docs/GCR2017-2018/05FullReport/TheGlobalCompetitivenessReport2017%E2%80%932018.pdf.
4. Anuradha Raman, "Weak Public Institutions Best Way to Ensure Social Injustice," *Hindu*, http://www.thehindu.com/todays-paper/tp-opinion/weak-public-institutions-best-way-to-ensure-social-injustice/article19361066.ece.
5. Karthik Muralidharan, "Priorities for Primary Education Policy in India's 12th Five-Year Plan," NCAER-Brookings India Policy Forum 2013, http://econweb.ucsd.edu/~kamurali/papers/Published%20Articles/Priorities%20for%20Primary%20Education%20Policy%20in%20India.pdf.
6. Mitra has explained his work with Hole in the Wall in several TED Talks, which I drew on to describe some of his experiments. Here are some of the

links to his videos: https://www.ted.com/talks/sugata_mitra_the_child_driven_education/transcript; https://www.ted.com/talks/sugata_mitra_shows_how_kids_teach_themselves.

7. Mark Warschauer, "The Paradoxical Future of Digital Learning," *Learning Inquiry* 1, no. 1 (2007): 41–49, http://education.uci.edu/uploads/7/2/7/6/72769947/paradox.pdf.
8. Bansal first described her "Chain of Hope" theory in a TED Talk in Paris in May 2016, https://www.ted.com/talks/seema_bansal_how_to_fix_a_broken_education_system_without_any_more_money.
9. Karthik Muralidharan, Abhijeet Singh, and Alejandro Ganimian, "Disrupting Education? Experimental Evidence on Technology-Aided Instruction in India," Working paper, 2017, last accessed February 9, 2018, http://econweb.ucsd.edu/~kamurali/papers/Working%20Papers/Disrupting%20Education%20(Current%20WP).pdf.

Chapter 3

1. I first reported on this story for CNN.com and have drawn on that reporting for this chapter: http://edition.cnn.com/2015/10/05/asia/india-jobs-crisis/.
2. Asia Pacific Human Development Report, "Shaping the Future: How Changing Demographics Can Power Human Development," United Nations Development Programme, 2016, last accessed February 9, 2018, http://hdr.undp.org/sites/default/files/rhdr2016-full-report-final-version1.pdf.
3. Press Information Bureau, Government of India, July 2, 2015, http://pib.nic.in/newsite/PrintRelease.aspx?relid=122929.
4. Lucas Chancel and Thomas Piketty, "Indian Income Inequality, 1922–2014: From British Raj to Billionaire Raj?" World Wealth and Income Database, http://wid.world/wp-content/uploads/2017/12/ChancelPiketty2017WIDworld.pdf.
5. Pranjul Bhandari and Prithviraj Srinivas, "More Jobs per Click," HSBC Global Research, 2016, http://www.hsbc.com/news-and-insight/insight-archive/2016/india-more-jobs-per-click.
6. I have drawn on the research of author and financial strategist Zachary Karabell for this section; Karabell has touched on this argument in his columns and also in his book *The Leading Indicators* (New York: Simon and Schuster, 2014).
7. Eric Brynjolfsson and JooHee Oh, "The Attention Economy: Measuring the Value of Free Goods on the Internet," MIT Initiative on the Digital Economy, http://ide.mit.edu/research-projects/attention-economy-measuring-value-free-goods-internet.
8. http://www.digitaldesh.in/.
9. https://hbr.org/product/zipdial-reaching-the-next-3-billion-consumers/IDE07-PDF-ENG.

Chapter 4

1. Sonalde Desai and Reeve Vanneman, *India Human Development Survey-II (IHDS-II), 2011–12* (Ann Arbor, MI: Inter-university Consortium for Political and Social Research [distributor], May 26, 2017), https://doi.org/10.3886/ICPSR36151.v5.
2. Snigdha Poonam, "Stalker's Delight: Mobile Numbers of Girls for Sale in UP Recharge Shops," *Hindustan Times*, February 3, 2017, http://www.hindustantimes.com/india-news/girls-mobile-numbers-up-for-sale-in-uttar-pradesh-price-rs-50-to-rs-500/story-5lYPcav12h7rnW6A6UDLLI.html.
3. Julia Qermezi Huang, "Digital Aspirations: 'Wrong Number' Mobile-Phone Relationships and Experimental Ethics among Women Entrepreneurs in Rural Bangladesh," *Journal of the Royal Anthropological Institute* 24, no. 1 (2018): 107–125.

Chapter 5

1. Google has a 96 percent share of the internet search market, so it is the most authoritative source on what Indians are searching for. Google Trends provides dynamic data on search patterns: https://trends.google.com/trends/topcharts#vm=cat&geo=IN&date=2012&cid.
2. Atish Srivastava, "Sunny Leone in 'Big Boss': BJP Youth Wing President Seeks Action against Colors," *India Today*, http://indiatoday.intoday.in/story/bjp-leader-tv-channel-sunny-leone-bigg-boss/1/162383.html.
3. Pornhub's 2016 Year in Review, https://www.pornhub.com/insights/2016-year-in-review.
4. Maureen O'Connor, "Pornhub Is the Kinsey Report of Our Time," *The Cut*, https://www.thecut.com/2017/06/pornhub-and-the-american-sexual-imagination.html.
5. Pornhub's 2017 Year in Review, https://www.pornhub.com/insights/2017-year-in-review.
6. Ananya Bhattacharya, "One of India's Busiest Train Stations Has Become a Porn Hub," QZ.com, https://qz.com/812470/indians-in-patna-are-using-googles-free-wifi-to-browse-porn/.
7. Nita Bhalla, "India Ministers Quit after Caught Watching Porn in Parliament," Reuters.com, http://in.reuters.com/article/us-india-porn-idINTRE 8170VD20120208.
8. https://twitter.com/SunnyLeone/status/898088217370664961.
9. Firstpost.com, "Wait, What? FIR Filed against Sunny Leone for Distribution of Obscene Content," http://www.firstpost.com/entertainment /wait-fir-filed-sunny-leone-distribution-obscene-content-2246776.html.
10. Zeenews.com, "FIR against Sunny Leone and Team 'Mastizaade' over Condom Scene," http://zeenews.india.com/entertainment/movies/fir-against-sunny-leone-and-team-mastizaade-over-condom-scene_1854174 .html.

11. CNN News18 is the Indian affiliate of CNN International. While the two networks share part of a name—a franchise-branding and content-sharing agreement—I had no part in operational planning of CNN News18's content while I was based in New Delhi as CNN International's bureau chief. I did sometimes appear on their programs as a guest.
12. Aamir Khan can be seen talking about Sunny Leone's CNN News18 interview on the following website: https://www.youtube.com/watch?v=OHz9x3VIqw8.
13. https://docs.google.com/document/d/1ZyBevXbdC-FXzkSNA9itU50FjhwO7CNSmZ7_HoJi_Bo/edit.
14. Anthony D'Amato, "Porn Up, Rape Down," Northwestern Public Law Research Paper No. 913013, https://papers.ssrn.com/sol3/papers.cfm?abstract_id=913013.

Chapter 6

1. "Even 70 Lakh Indian Army Cannot Defeat Kashmiris: Arundhati Roy," *Times of Islamabad,* May 16, 2016, https://timesofislamabad.com/16-May-2017/even-70-lakh-indian-army-cannot-defeat-kashmiris-arundhati-rai.
2. Jean M. Twenge, "Have Smartphones Destroyed a Generation?" *Atlantic,* https://www.theatlantic.com/magazine/archive/2017/09/has-the-smartphone-destroyed-a-generation/534198/.
3. David Streitfeld, "The Internet Is Broken: @ev Is Trying to Salvage It," *New York Times,* https://www.nytimes.com/2017/05/20/technology/evan-williams-medium-twitter-internet.html.
4. Olivia Solon, "Ex-Facebook President Sean Parker: Site Made to Exploit Human 'Vulnerability,'" *Guardian,* https://www.theguardian.com/technology/2017/nov/09/facebook-sean-parker-vulnerability-brain-psychology.
5. Julia Carrie Wong, "Former Facebook Executive: Social Media Is Ripping Society Apart," *Guardian,* https://www.theguardian.com/technology/2017/dec/11/facebook-former-executive-ripping-society-apart.
6. https://www.facebook.com/zuck/posts/10104074437830721.
7. Isa Soares and Florence Davey-Attlee, "The Fake News Machine: Inside a Town Gearing Up for 2020," CNNMoney.com, http://money.cnn.com/interactive/media/the-macedonia-story.
8. Art Swift, "Americans' Trust in Mass Media Sinks to New Low," Gallup News, http://news.gallup.com/poll/195542/americans-trust-mass-media-sinks-new-low.aspx.
9. Nicholas Carr, "Is Google Making Us Stupid?" *Atlantic,* https://www.theatlantic.com/magazine/archive/2008/07/is-google-making-us-stupid/306868/.
10. Roger McNamee, "How to Fix Facebook Before It Fixes Us," *Washington Monthly,* https://washingtonmonthly.com/magazine/january-february-march-2018/how-to-fix-facebook-before-it-fixes-us/.

Chapter 7

1. Human Rights Watch, "India: 20 Internet Shutdowns in 2017," https://www.hrw.org/news/2017/06/15/india-20-internet-shutdowns-2017.
2. Darrell M. West, "Internet Shutdowns Cost Countries $2.4 Billion Last Year," Center for Technology Innovation at Brookings, https://www.brookings.edu/wp-content/uploads/2016/10/intenet-shutdowns-v-3.pdf.
3. https://www.medianama.com/wp-content/uploads/Rules-Temporary-Suspension-of-Telecom-Services-Internet-Shutdowns-Aug-2017.pdf.
4. Leo Mirani, "Millions of Facebook Users Have No Idea They're Using the Internet," QZ.com, https://qz.com/333313/milliions-of-facebook-users-have-no-idea-theyre-using-the-internet/.
5. https://www.globe.com.ph/press-room/globe-extends-free-facebook.
6. Osama Manzar, "Zuckerberg, Facebook, Internet.org and Indian Villages," LiveMint.com, http://www.livemint.com/Opinion/FBePf7xWFjQ3rxyQ3UDTBI/Zuckerberg-Facebook-Internetorg-and-Indian-villages.html.
7. "The Global Economic Impact of Facebook: Helping to Unlock New Opportunities," Deloitte, https://www2.deloitte.com/uk/en/pages/technology-media-and-telecommunications/articles/the-global-economic-impact-of-facebook.html.
8. https://www.facebook.com/narendramodi/.
9. Nikhil Pahwa, "What Mark Zuckerberg Didn't Say about Internet.org," Medianama.com, https://www.medianama.com/2014/10/223-zuckerberg-india-internet-org/.
10. https://www.youtube.com/watch?v=mfY1NKrzqio.
11. https://www.theguardian.com/technology/2016/may/12/facebook-free-basics-india-zuckerberg.
12. Mark Zuckerberg, "Free Basics Protects Net Neutrality," *Times of India*, https://blogs.timesofindia.indiatimes.com/toi-edit-page/free-basics-protects-net-neutrality/.
13. Nikhil Pahwa, "It's a Battle for Internet Freedom," *Times of India*, https://blogs.timesofindia.indiatimes.com/toi-edit-page/its-a-battle-for-internet-freedom/.
14. I drew on my reporting for CNN.com in writing this section: Ravi Agrawal, "India Supreme Court Rules Privacy 'Fundamental Right' in Landmark Case," CNN.com, http://www.cnn.com/2017/08/24/asia/indian-court-right-to-privacy/index.html.
15. Siddharth George and Arvind Subramanium, "Transforming the Fight against Poverty in India," *New York Times*, July 22, 2015, https://www.nytimes.com/2015/07/23/opinion/transforming-the-fight-against-poverty-in-india.html.
16. Rachna Khaira, "Rs 500, 10 Minutes, and You Have Access to Billion Aadhaar Details," *Tribune*, January 4, 2018, http://www.tribuneindia.com/news/nation/rs-500-10-minutes-and-you-have-access-to-billion-aadhaar-details/523361.html.

17. https://twitter.com/snowden/status/950490382990790656.

Chapter 8

1. I first interviewed Sarvesh Kumar in my reporting for CNN: https://www.cnn.com/2016/11/17/asia/india-rickshaw-rupee/index.html.
2. I first interviewed Paytm's Vijay Shekhar Sharma for CNN: https://www.cnn.com/videos/world/2016/11/30/inside-india-cash-crunch-pkg-agrawal.cnn.
3. Anant Vijay Kala, "5 Ways India's Cash Squeeze Is Going to Affect the Economy," *Wall Street Journal*, November 16, 2016, https://blogs.wsj.com/briefly/2016/11/16/5-ways-indias-cash-squeeze-is-going-to-affect-the-economy/.
4. BS Web Team, "RBI Annual Report: 99% of Demonetized Currency Back with Central Bank," *Business Standard*, August 30, 2017, http://www.business-standard.com/article/economy-policy/rbi-annual-report-99-of-demonetised-currency-back-into-the-system-117083000891_1.html.
5. Kaushik Basu, "In India, Black Money Makes for Bad Policy," *New York Times*, November 27, 2016, https://www.nytimes.com/2016/11/27/opinion/in-india-black-money-makes-for-bad-policy.html.
6. Shailesh Kumar's quote to me was first used in a report for CNN: https://www.cnn.com/2017/03/11/asia/uttar-pradesh-elections/index.html.

Conclusion

1. The launch of Apple's iPhone has been widely covered in the media. While I was not in attendance at the conference, I drew on the recorded video of the event for this reporting: https://www.youtube.com/watch?v=x7qPAY9JqE4.
2. Rishi Iyengar, "An iPhone Now Costs $1,700 in India," CNNMoney.com, February 5, 2018, http://money.cnn.com/2018/02/05/technology/apple-iphone-india-price-increase-tariffs/index.html.
3. Farhad Manjoo, "Why Tech Is Starting to Make Me Uneasy," *New York Times*, October 11, 2017, https://www.nytimes.com/2017/10/11/insider/tech-column-dread.html.
4. Jenny Anderson, "A Letter from Two Big Apple Investors Powerfully Summarizes How Smartphones Mess with Kids' Brains," *Quartz*, January 8, 2018, https://qz.com/1174317/a-letter-from-apple-aapl-investors-jana-partners-and-calstrs-powerfully-summarizes-how-smartphones-mess-with-kids-brains/.
5. For parts of this section I have drawn on reporting from Gurcharan Das, *India Unbound* (New Delhi: Penguin India, 2001).
6. Shashi Tharoor, "Meanwhile: India's Cellphone Revolution," *New York Times*, February 2, 2007, http://www.nytimes.com/2007/02/01/opinion/01iht-edtharoor.4431582.html.
7. Rajeev Mantri and Harsh Gupta, "The Story of India's Telecom Revolution," Livemint.com, January 8, 2013, http://www.livemint.com/Opinion/biNfQImaeobXxOPV6pFxqI/The-story-of-Indias-telecom-revolution.html.

8. Arvind Panagariya, *India: The Emerging Giant* (New Delhi: Oxford University Press, 2008).
9. Charles Riley and Medhavi Arora, "India's Richest Man Offers Free 4G to One Billion People," CNNMoney.com, September 6, 2016, http://money.cnn.com/2016/09/06/technology/india-reliance-jio-4g-internet/index.html.
10. Rishi Iyengar, "India's Mobile Price War Just Claimed Another Victim," CNNMoney.com, February 28, 2018, http://money.cnn.com/2018/02/28/technology/aircel-bankruptcy-india-mobile-price-war/index.html.
11. Rishi Iyengar, "A Huge Wave of New Users Is Killing 4G Speeds in India," CNNMoney.com, June 8, 2017, http://money.cnn.com/2017/06/08/technology/india-4g-reliance-jio-speed/index.html.
12. Tiffany Hsu, "Alibaba's Singles Day Sales Hit New Record of $25.3 Billion," *New York Times*, November 10, 2017, https://www.nytimes.com/2017/11/10/business/alibaba-singles-day.html.
13. I have touched on these trends in my reporting for CNN over the years. One such essay that I drew on for this section is at https://www.cnn.com/2017/08/08/opinions/independence-ravi-agrawal-opinion/index.html.
14. World Population Prospects 2017, Population Division, Department of Economic and Social Affairs, United Nations Secretariat, https://esa.un.org/unpd/wpp/Graphs/.
15. World Bank data on life expectancy, https://www.google.com/publicdata/explore?ds=d5bncppjof8f9_&met_y=sp_dyn_le00_in&idim=country:IND.
16. "The Future of World Religions: Population Growth Projections 2010–2050," Pew Research Center, April 2, 2015, http://www.pewforum.org/2015/04/02/religious-projections-2010-2050/.

SELECTED BIBLIOGRAPHY

Here are some of the books I have drawn on, learned from, or made references to at various points. This is not an exhaustive list. The authors are listed in alphabetical order. Each of these is a great read and could be useful for further research on the themes and trends I covered in *India Connected.*

Brynjolfsson, Erik, and Andrew McAfee. *The Second Machine Age: Work, Progress, and Prosperity in a Time of Brilliant Technologies.* New York: W. W. Norton, 2014.

Bumiller, Elisabeth. *May You Be the Mother of a Hundred Sons: A Journey among the Women of India.* New York: Ballantine Books, 1991.

Carr, Nicholas. *The Shallows: How the Internet Is Changing the Way We Think, Read, and Remember.* New York: W. W. Norton, 2010.

Das, Gurcharan. *India Grows at Night.* New Delhi: Penguin India, 2013.

Das, Gurcharan. *India Unbound.* New Delhi: Penguin India, 2001.

Doron, Assa, and Robin Jeffrey. *The Great Indian Phonebook: How the Cheap Cell Phone Changes Business, Politics, and Daily Life.* Cambridge, MA: Harvard University Press, 2013.

Foer, Franklin. *World Without Mind: The Existential Threat of Big Tech.* New York: Jonathan Cape, 2017.

Fukuyama, Francis. *Trust: The Social Virtues and the Creation of Prosperity.* New York: Simon & Schuster, 1996.

Giridharadas, Anand. *India Calling: An Intimate Portrait of a Nation's Remaking.* New York: Times Books, 2011.

Guha, Ramachandra. *India After Gandhi: The History of the World's Largest Democracy.* New Delhi: Pan Macmillan India, 2017.

Kardaras, Nicholas. *Glow Kids: How Screen Addiction Is Hijacking Our Kids—and How to Break the Trance.* New York: Macmillan, 2016.

Khilnani, Sunil. *The Idea of India.* New York: Farrar, Straus & Giroux, 1999.

Luce, Edward. *In Spite of the Gods: The Rise of Modern India.* New York: Anchor, 2008.

Mele, Nicco. *The End of Big: How the Internet Makes David the New Goliath.* New York: St. Martin's Press, 2013.

Merchant, Brian. *The One Device: The Secret History of the iPhone.* New York: Little, Brown, 2017.

Morozov, Evgeny. *The Net Delusion: The Dark Side of Internet Freedom.* New York: PublicAffairs, 2011.

Naipaul, V. S. *India: A Wounded Civilization.* New York: Vintage Books, 2003.

Nilekani, Nandan. *Imagining India: Ideas for the New Century.* New Delhi: Penguin India, 2010.

Nilekani, Nandan, and Viral Shah. *Rebooting India: Realizing a Billion Aspirations.* New Delhi: Penguin India, 2015.

Panagariya, Arvind. *India: The Emerging Giant.* New Delhi: Oxford University Press, 2008.

Pitroda, Sam. *Dreaming Big: My Journey to Connect India.* New Delhi: Penguin India, 2015.

Poonam, Snigdha. *Dreamers: How Young Indians Are Changing the World.* New Delhi: Penguin Random House, 2018.

Raghunathan, V. *Games Indians Play: Why We Are the Way We Are.* New Delhi: Penguin India, 2007.

Sengupta, Somini. *The End of Karma: Hope and Fury among India's Young.* New York: W. W. Norton, 2016.

Sharma, Ruchir. *Breakout Nations: In Pursuit of the Next Economic Miracles.* New York: W. W. Norton, 2012.

Taplin, Jonathan. *Move Fast and Break Things: How Facebook, Google, and Amazon Cornered Culture and Undermined Democracy.* New York: Little, Brown, 2017.

Tharoor, Shashi. *The Elephant, the Tiger, and the Cellphone: India, the Emerging 21st Century Power.* New York: Arcade, 2011.

Trivedi, Ira. *India in Love: Marriage and Sexuality in the 21st Century.* New Delhi: Aleph, 2014.

Twenge, Jean M. *iGen: Why Today's Super-Connected Kids Are Growing Up Less Rebellious, More Tolerant, Less Happy—and Completely Unprepared for Adulthood.* New York: Atria Books, 2017.

Wu, Tim. *The Attention Merchants: The Epic Scramble to Get Inside Our Heads.* New York: Knopf, 2016.

Zakaria, Fareed. *The Post-American World.* New York: W. W. Norton, 2008.

INDEX

A

Aadhaar, 12–13, 59, 166–67, 202
addiction, 136–38
adult films, 126–27. *See also* pornography
advertising, 15, 143, 199–201
Ahmed, Mukhtar, 149–58
Airtel, 81, 149, 154, 182, 197, 198, 203
Akhlaq, Mohammad, 133
Akhtar, Faheem, 53
Alibaba, 78
All India Bakchod (AIB), 164
al-Qaeda, 152–53
AltNews, 130
Amazon, 16, 78, 80
Ambani, Mukesh, 198
American Dream, 14
Anandan, Rajan, 23
Anantnag, 147–65
Android-based phones, 4
anganwadis, 62–63
Apple, 4, 16, 191–94
apps, 82–85, 92–93, 178–81. See also *specific apps*
Arora, Simran, 17
arranged marriage, 89–90
AskMonk, 99–100
astrology, 98–100
Atlantic, 138, 142
The Attention Merchants: The Epic Scramble to Get Inside Our Heads (Wu), 141

B

Babajob, 83–84
"Baby Doll" (music video), 116
Bahl, Kunal, 77–78
Bansal, Seema, 61–63
Basu, Jyoti, 13, 186
Basu, Kaushik, 10
BCG (Boston Consulting Group), 61–62
Bedi, Kiran, 124–25
Bezos, Jeff, 16
Bhandari, Pranjul, 77–79
Bhandari, Pranshu, 54–57
Bhatia, Rahul, 164
Bhatia, Sachin, 93–98
Bhatt, Alia, 41–42
Bhatt, Mahesh, 116
Big Boss, 115–16
Bing, 79
bit cinema, 127

Black Mirror (Netflix series), 194
black money, 174
Blagsvedt, Sean, 82–84
blue films, 127
BookMyChotu, 176–77
Boston Consulting Group (BCG), 61–62
Brazil, 117
Brexit, 141
bribery, 65–66
Brynjolfsson, Erik, 76, 79
BuzzFeed, 141–42
B2X Consumer Trends, 139

C
cable television, 13–14
California State Teachers' Retirement System, 194
Call of Duty (videogame), 136
Carr, Nicholas, 142–43
cars, 5–6, 69–70
Centre for Social Research (CSR), 126
Chadha, Sapna, 35
chain emails, 131–32
The Chain of Hope, 62
Chancel, Lucas, 71
Charlie and the Chocolate Factory (film), 135
Chaubey, Bhupendra, 119–21
China, 16, 138
Chopra, Honey, 98–100, 103, 108–9
Christian weddings, 103–4
cinema, bit, 127
Clinton, Hillary, 141, 171
CNN, 119, 166, 171
coaching centers, 44–46
Coates, Ta-Nehisi, 51
commercials, 199–201
connectivity, 195–96
Cook, Tim, 16
corruption, 10, 11, 65–66, 177–78, 185
CSR (Centre for Social Research), 126
cultural developments, 14–15
Culture Alley, 54
cybercafés, 150

D
Dahl, Roald, 135
D'Amato, Anthony, 125
Dangal (film), 124
Das, Gurcharan, 10, 195
dating, 89–113
  astrology and, 98–100
  illustrated story of, 89–93, 100–103, 106–13
  and Truly Madly Deeply, 93–98
  and weddings, 103–6
dating apps, 92–93
"death by selfie," 12
demonetization, 169–74, 175–78
Development as Freedom (Sen), 84
Didi Chuxing, 70
digital blackouts. *See* internet shutdowns
Digital Desh (Gupta), 79
Digital India, 26, 162, 178
digital money, 169–89
  aftermath of, 184–88
  and demonetization, 169–74, 175–78
  growth of, 178–83
  and tax avoidance, 174–75
  trust of, 183–84
Doordarshan, 13
driver incentives, for Uber, 72–74
Dunzo, 83–84

E
East India Company, 49
e-commerce, 7, 75–79, 197
economic optimization, 10
education, 43–63
  Abdul Wahid on effects in, 43–49
  and EkStep, 58–60, 62–63
  history of, 49–50
  and Hole in the Wall, 60–62
  and learning English, 50–57
  problems with, 57–58
  sex, 126–28
EkStep, 58–60
English, learning, 4–5, 50–57

F
Facebook, 121
  banning of, 153, 158
  as free service, 79
  and Internet.org, 159–65, 168
  and nomophobia, 140–41
  and WhatsApp, 16, 131
fake news, 132–33, 141, 142
fear of missing out (FOMO), 142
FIFA World Cup (2010), 85
films, 126–27. *See also* pornography
First Machine Age, 76
Flipkart, 78
Foer, Franklin, 142
FOMO (fear of missing out), 142
Ford, Henry, 193
4G technology, 198
Free Basics, 165
free services, 79–81
Fukuyama, Francis, 9

G
Gallup, 142
Games Indians Play: Why We Are the Way We Are (Raghunathan), 9
Gandhi, Indira, 195–96
Gandhi, Rahul, 166
Ganesha, 130
Ganguly, Meenakshi, 148
Ganimian, Alejandro, 63
Gender Gap Index, 39
Germany, 66
Gionee, 4
Global Competitiveness Index, 11, 57
Globe, 160
Glow Kids (Kardaras), 138
Goods and Services Tax (GST), 147, 149, 152–53
Google, 21–25, 35–42, 79, 118
Grindr, 94
GST (Goods and Services Tax), 147, 149, 152–53
"Gujarat Miracle," 28
Gupta, Ishaan, 79–80

H
Hajra, Sujan, 185
Haleem, Sheikh, 129–30
Hanuman, 22
Hello English, 50, 53–57
"high-trust" societies, 9
Hindi-medium schools, 51
Hindustan Times, 96
Hindu weddings, 103–4
Hole in the Wall, 60–62
hope, internet as symbol of, 35–36
The Hot Seat (talk show), 119
HSBC, 16, 77–78
Huang, Juli Q., 97
Huawei, 4

I
Ibtada, 23–24
Idea Cellular, 198, 199
iGen, 129–46
  and fake news, 132–33, 141–42
  and smartphone addiction, 136–38
IIP (Is it Possible), 43–49, 51
IITs (Indian Institutes of Technology), 45, 72
incentives, for Uber, 72–74
India Human Development Survey (2012), 89
India in Love (Trivedi), 126, 128
Indian Institutes of Technology (IITs), 45, 72
India Unbound (Das), 195
innovation, 77
Instagram, 153, 168
International Monetary Fund, 13
internet
  access to, 3–4
  addiction to (*See* nomophobia)
  data for, 55
  effects of, on family, 42
  shutdowns of (*See* internet shutdowns)
Internet Addiction Disorder, 138. *See also* nomophobia
Internet.org, 160–65

internet service providers (ISPs), 149
internet shutdowns, 12, 147–68
  in Anantnag, 147–65
  and privacy, 165–68
iPhones, 4, 192–93
Iraq, 148
ISPs (internet service providers), 149

J
Jain, Amit, 72–74
Jain, Pankaj, 133–35
Jaitley, Arun, 185
JAM, 166–67
JANA Partners LLC, 194
Japan, 66
Jharkhand, 129–32
Jio, 15–16, 132, 149, 154, 162, 198
Jivrajka, Pranay, 10, 72, 74
job creation, 65–85
  apps and, 82–85
  e-commerce and, 75–79
  and free services, 79–81
  issues in, 65–67
  and missed calls, 81–82
  and Uber, 67–75
Jobs, Steve, 52, 191–92, 193

K
Kalanick, Travis, 69–70
Kalidasa, 127
Kāmadeva, Lord, 90
Kama Sutra, 127–28
Kantar, 139
Kapur, Devesh, 57–58
Karbonn, 4
Kardaras, Nicholas, 138
Kashbook, 158–59, 168
Kashmir, 12
Keynes, John Maynard, 76
Khaira, Rachna, 167
Khan, Aamir, 121, 124
Khan, Sal, 52
Khan Academy, 52
Khan Market, 2–3
Khera, Shiv, 52
kiranawallahs, 182
Kolkata, 8, 10, 13
Kothari, Prakash, 125
Kranzberg, Melvin, 194
kshatriya, 31
Kumar, Asha Ram, 170, 188
Kumar, Babloo, 76, 201
Kumar, Rahul, 94
Kumar, Sarvesh, 169–70, 176–77, 186–89, 202
Kumar, Shailesh, 186
Kumari, Ranjana, 126

L
Lava, 4
The Laws of Manu, 128
learning English, 4–5, 50–57
LinkedIn, 83
literacy, 23, 57
lottery system, 63
"low-trust" societies, 9
LVMH, 174

M
Macaulay, Thomas Babington, 49–50
Macintosh, 191–92
Macworld trade show, 191–92
Madison, James, 165–66
Magon, Vaibhav, 99–100
MakeMyTrip.com, 94
Manjoo, Farhad, 194
Manzar, Osama, 40, 161, 163
marriage, 89–90, 103–8
Maruwada, Shankar, 59
Marwaris, 56
Marx, Karl, 15, 182
mass media, 142–43, 171
mass piracy. *See* side-loading
Masters, William, 125–26
Mastizaade (film), 119

Maurya, Deepanshu, 80–81
McAfee, Andrew, 76
McNamee, Roger, 143
media, 142–43, 171
MediaNama, 163
Mehsana, Gujarat, 25–35
melia, 31
Merchant, Brian, 192
Micromax, 4, 197
Mill, John Stuart, 166
Ministry of Communications, 122
Mint (newspaper), 161
misogyny, 34–35
missed call, 81–82
Mitra, Sugata, 60–61
Mobile Marketing, 139
Model T, 193
Modi, Narendra
  and demonetization, 171–73, 178
  influence of, 28, 133
  and Internet.org, 161–62
  and missed calls, 82
money, 174. *See also* digital money
MoneyOnMobile, 181–84, 202
Mostly Sunny (documentary), 119, 124–25
Muralidharan, Karthik, 58–60, 63

N
Nangli Jamawat, 21–25, 35–36
Narayan, R. K., 17
National Crime Records Bureau, 38
National Institute of Information Technology (NIIT), 60
nationalism, 14
National Skill Development Mission, 66
Nehru, Jawarharlal, 126–27
Netflix, 194
net neutrality, 162–65
New Delhi, 2–3, 13
New Republic, 142
New York Times, 140, 185
NIIT (National Institute of Information Technology), 60
Nilekani, Nandan, 8, 59
Nilekani, Rohini, 59
Nirbhaya: Be Fearless (app), 38
nomophobia, 12, 136–40

O
Ola, 7, 10, 67, 70–72
Oliver, John, 164
The One Device (Merchant), 192
One Night Stand (film), 122
Oppo, 4

P
Pahwa, Nikhil, 163–66, 168
Palihapitiya, Chamath, 141
Parched (film), 26–27
Parker, Sean, 140
Patni, Nishant, 54, 56–57
Paytm, 178–81
"peons," 65–66
Peretti, Jonah, 141
Philippines, 117
Piketty, Thomas, 71
piracy, mass. *See* side-loading
Pitroda, Sam, 195
Playboy, 126–27
political parties, 9
Poonam, Snigdha, 96–97
Pornhub, 116–18
pornography, 115–28
  and ascetic side of Indian culture, 128
  effect of, 12, 201
  history of access to, 126–27
  and Sunny Leone, 118–24
  and rape, 124–26
  and sex education, 126
  viewing of, 117–18
"Porn Up, Rape Down" (D'Amato), 125
poverty, 4, 7–8
Prasad, Ravi Shankar, 166

Prasad, Sudhir, 16–17, 72–74, 201
prepaid cellular, 82
Press Trust of India, 25–26
privacy, 41, 93, 127, 165–68, 202
propaganda, 15
public trust, 9–11
punctuality, 10

Q
Quartz, 160

R
Raghunathan, V., 9
Raghuvamsa (Kalidasa), 127
Rajasthan, 1–2
Ramayana, 33
rape, 37–38, 124–26
Rawal, Paresh, 134
relationships, wrong number, 97
Reliance Industries, 4, 15–16, 132, 198
religion, 204
Reserve Bank of India, 185
Roka, 107
Rowling, J. K., 192
Roy, Arundhati, 134
Roy, Aviroop, 183–84

S
Saathi, 21–42
Sadhwani, Mohit, 3
Safetipin, 38–39
Sahni, Manoj Kumar, 23–24
Sandberg, Sheryl, 159–60
sarpanch, 25–26
Savage Garden (band), 94
Savita Bhabhi (comic strip), 123
Second Machine Age, 76–77
The Second Machine Age: Work, Progress, and Prosperity in a Time of Brilliant Technologies (Brynjolfsson and McAfee), 76
Sen, Amartya, 84
senior citizens, 204
Service for Healthy Use of Technology (SHUT), 137
sex education, 126–28
Shaadi.com, 92
Shafiq, Zeyan, 147–48, 158–59, 168, 202
Shaikh, Mohsin Sadiq, 133
Sharma, Manoj Kumar, 137–38, 140
Sharma, Toshi, 53
Sharma, Vijay Shekhar, 178–81
Shaw, Guru Prasad, 8, 182
Shaw, Sanjay, 8, 182–83
Shaw, Sunil, 8, 182
SHUT (Service for Healthy Use of Technology), 137
side-loading, 11
Singh, Abhijeet, 63
Sinha, Pratik, 130–31, 133–34
Sinha, Saikat, 17, 136–37, 139–40, 201
smartphone addiction, 136–46
  effects of, 139–40, 201
  prevalence of, 12, 17
  and social media, 140–41
  treatment of, 137–38
smartphones
  addiction to (*See* smartphone addiction)
  Android-based, 4
  ban of, 25–35, 48
  current state of, 202–3
  development of, 3
  effects of, 6–7, 17
  and empowerment of women, 35–42
  events leading to adoption of, 13–14
  future of, 203–5
  limitations of, 8
  resistance to, 21–25
  uses of, 5–6, 194
  weaponized, 11–12
  and women's safety, 38–39
SMHoaxSlayer, 133–34
Snapdeal, 78
Snowden, Edward, 167
Soares, Isa, 141
social media, 143
societies, high-trust vs. low trust, 9
Sofi, Mushtaq Ahmed, 155–56

Software Freedom Law Centre, 148
Solangi, Jagat Singh, 29–30
Sony, 192
South Africa, 117
South Korea, 66, 138
Stephen, C. M., 195
Sunny Leone, 115–28
and Hot Seat interview, 119–21
interview with, 121–24
popularity of, 118–19
rise of, 115–16
Supreme Court, 165–66
Suraj (village), 25–35
Syria, 148

T
Taj Mahal, 2
Taneja, Vineet, 197
Taobao villages, 78–79
Tata Trusts, 22
tax avoidance, 174–75, 177
technological unemployment, 76
technology, 15–16, 23
TED Talks, 52
Telecom Regulatory Authority of India (TRAI), 163–65
Telenor, 198
television, cable, 13–14
Tewari, Abhishek, 175
Thakore, Ramaji, 30–34
Thakur, Anurag, 115–16
Times of Islamabad, 134
Tinder, 92–93, 100
TM Spark, 97
Tokyo, Japan, 9–10
Toyota, 192
traditional weddings, 103–6
TRAI (Telecom Regulatory Authority of India), 163–65
Transparency International, 11
Tribune, 167
Trivedi, Ira, 126, 128
Truly Madly (app), 93–98, 100–101
"Truly Madly Deeply" (song), 94
Trump, Donald, 141, 171
trust, 9–11, 95, 183–84
tutoring, 44–46
Twenge, Jean, 138–39
Twitter, 85, 121, 140, 153, 158
2G spectrum scam, 197

U
Uber, 7, 10, 39, 67–75, 177
UNDP (United Nations Development Programme), 66
unemployment, technological, 76
Unique Identification Authority of India, 167
United Kingdom, 66
United Nations Development Programme (UNDP), 66
United States, 5–6, 66, 103–4
UP (Utter Pradesh), 65
UrbanClap, 67
US election (2016), 141

V
Vankar, Devshi, 27–28
vari, 31
Vidyalakshmi Bhavan school, 34
Vodafone, 149, 154, 182, 198, 199
Vohra, Karenjit Kaur. *See* Sunny Leone
voice technology, 23
Vulcan, 75–76

W
Wagoner, Valerie, 85
Wahid, Abdul, 16, 43–49, 51–53, 201, 205
Wall Street Journal, 185
Wani, Burhan, 150–51
The War of the Worlds (Wells), 133
Warschauer, Mark, 61
Washington Monthly, 143
Weber, Daniel, 115
weddings, traditional, 103–6
Welles, Orson, 133
Wells, H. G., 133
West Bengal, 13

WhatsApp
banning of, 158
and digital money, 181
and education, 62
and Facebook, 16, 131
and fake news, 132–35
and job creation, 80
and nomophobia, 129–35
white money, 174
Wi-Fi, 55
Wikipedia, 79
Williams, Evan, 140
women, 21–42
ban on smartphones for, 25–35
challenges for, 39
and dating, 95–96
empowerment of, 40
exposure to smartphones, 24–25
and rape, 38
smartphone and empowerment of, 35–42
smartphone use by, 22
tasks of, 37
World Economic Forum, 11, 39, 57, 159–60
World Without Mind (Foer), 142
wrong number relationships, 97
Wu, Tim, 141

X
Xiaomi, 4, 16

Y
Yadav, Pankaj, 28–29, 31–32
You Can Win (Khera), 52
YouTube, 42, 55, 79, 116

Z
ZipDial, 85
Zuckerberg, Mark, 141, 160–65, 168